T0135622

PhD thesis

The evolution of genetic representations and modular neural adaptation

Dissertation zur Erlangung des Grades eines Doktors der
Naturwissenschaften in der Fakultät für Physik und Astronomie der
Ruhr-Universität Bochum

Marc Toussaint

*Institut für Neuroinformatik, Ruhr-Universität Bochum, ND 04,
44780 Bochum—Germany*

mt@neuroinformatik.ruhr-uni-bochum.de

March 31, 2003

Bibliografische Information Der Deutschen Bibliothek
Die Deutsche Bibliothek verzeichnet diese Publikation in der Deutschen
Nationalbibliografie; detaillierte bibliografische Daten sind im Internet über
http://dnb.ddb.de abrufbar.

ISBN 3-8325-0579-2

Logos Verlag Berlin
Comeniushof, Gubener Str. 47,
10243 Berlin
Tel.: +49 030 42 85 10 90
Fax: +49 030 42 85 10 92
INTERNET: http://www.logos-verlag.de

Gutachter: Prof. Dr.-Ing. Werner von Seelen, Prof. Dr. Klaus Goeke

Contents

Introduction

A matter of representation. When trying to understand biological systems, a first step is typically descriptive, i.e., aiming at a precise description of a system's functionalities and mechanisms. This is true for many branches of biology when describing the subtle mechanisms of natural life, for many branches of Neuroscience describing in detail neural mechanisms, and eventually also for classical approaches to Artificial Intelligence trying to capture intelligence on a formal descriptive basis. However, a purely descriptive approach might neglect that *all* the intelligent systems we find in nature are the outcome of adaptation processes, namely evolution and neural adaptation. This puts a crucial constraint on what systems can possibly exist in nature, in particular how they must be organized and structured to allow for adaptation. This calls for a theory on these adaptation processes themselves as the basis to understand natural intelligent systems.

Why does the need for adaptation induce a structural constraint? The point is that nature does not invent intelligent systems directly. All of a system's functionality is represented *indirectly* by genes or neurons and adaptation takes place on the level of these system parameters. The way these parameters encode the system's final features is highly non-trivial. For example, engineers often profit from precise descriptions of nature and have adopted many techniques. However, when they design systems they usually directly describe and design the system's final functionality; they hardly adopt nature's strategy to use an indirect representation of functionalities by genes or some equivalent. This leads to a problem that becomes particularly apparent when the goal is the design of an Artificial Intelligence. Early approaches to Artificial Intelligence were largely descriptive, i.e., tried to first directly describe what intelligent behavior is in terms of behavioral rules. These rules could then be implemented on a computer. But actually, a precise *direct* description of intelligence in terms of rules is very difficult. One realizes that whenever one formulates a behavioral rule, one needs to presume some vocabulary of situations and behaviors to formulate the rule (e.g., "when in situation A execute behavior b" presumes that it is well-defined what situation A and behavior b are). Nature developed a elaborate representation to encode intelligent behavior, namely highly structured neural systems. This representation is the outcome of a long adaptive process.

The crucial question is how the adaptation processes succeeded in developing these indirect representations that are so beneficial for functional adaptation.

Complex adaptation mechanisms on arbitrary representations— or simple adaptation mechanisms on suitable representations?

Research in the field of Artificial Intelligence has developed more and more sophisticated adaptation mechanisms that mimic natural evolution and learning. For instance, a breakthrough was certainly the idea to use gradient learning for all kinds of neural models with real-valued parameters. This adaptation mechanism proved to be a generic tool and theories on the universal approximation capabilities of multi-layer feed-forward neural networks show that artificial neural networks can, in principle, represent (or approximate) any functionality. At first sight it seems that the adaptation problem is solved and thus there is little motivation to reconsider the *way* conventional artificial neural networks represent functionalities—instead one would praise gradient learning to be in some sense independent of how functionalities are represented. A simple example though shows how important it is to consider how functionalities are represented: When *we* learn something we almost consciously know that it will affect only that very functionality that we intend to learn. This can be possible only when the functionality is represented in such a way that we can adapt it in one step and without worrying that other functionalities of our brain are affected. Artificial neural networks represent functionalities in a way such that, if one functionality is trained, it is highly likely that many other functionalities are affected—presumably in a negative way (cf. the credit assignment problem, catastrophic forgetting, cross-talk, etc.). As a consequence, supervised stochastic online learning in artificial neural networks is a long process where every functionality has to be learned effectively simultaneously by switching between the tasks all the time and adapting only in small steps. But instead of rethinking the way functionalities should be represented, research focuses on further developing the adaptation mechanisms. Gradient learning is already far more complex than local learning rules like the Hebb rule and its further developments (Rprop, natural gradient, conjugate gradient, BFGS, etc.) become more and more complex.

Nature exhibits quite the opposite strategy to solve the dilemma. Compared to the adaptation mechanisms developed in Artificial Intelligence research, the principles of natural adaptation mechanisms seem rather elementary. Neuroscience tells us that the principles of neural plasticity are not that far from the classical Hebb rule on the level of synaptic plasticity. And the basic principle of evolutionary adaptation is the trial-and-error strategy on the level of gene mutations. In view of the simplicity of principles of these adaptation mechanisms—*on the level of synapses and genes*—the capability of complex adaptation on the functional level is astonishing. The reason must be a suitable choice of how functionalities are represented. And here, nature found very complex solutions. A gene mutation affects the functional traits of an organism via complex processes of gene expression during ontogenetic development. Although the human genome project succeeded in enumerating all the genes, we are far from knowing what these genes actually affect; the networks of gene interaction are too complex, but obviously not random and have been subject to evolutionary adaptation. In the brain, behavior is the result of complex processes of recurrent activation dynamics. For higher-level brain areas we do not know how synapses store our knowledge and how they affect the global activation dynamics when we read out this knowledge.

Nature developed these representations in the course of adaptive processes and it thus should be a general project to develop theories on adaptation processes that account for the adaptation of representations. In its main chapter, this thesis will offer an integrative perspective and theory on the adaptation of genetic representations which turns out to be inherent in every ordinary evolutionary process with non-trivial genotype-phenotype mapping. It will discuss what "suitable" means and show that genetic representations indeed evolve as to optimize their suitability. In the second chapter we also, more briefly, address the case of neural adaptation. Here, a major result is the proof that representations in conventional artificial neural networks are indeed uncontrolled in the sense we mentioned it. In such systems it is highly unlikely that representations are structured and organized such that different functionalities can be learned without cross-affecting each other.

A formal framework for structured adaptation. The theoretical approach pursued in this thesis to treat the issue of suitable and adaptive representations is based on the a specific aspect of general adaptational processes, namely the aspect of correlational structure in adaptation steps. Let us make this concept more explicit which both, the chapters on evolutionary and neural adaptation, share.

Consider a system which can acquire different states in a state space X; we enumerate these states by i. We describe this system not by its current state but by the current probability field $\psi(t) \in \Lambda^X$ over states. Here, Λ^X is the space of all probability distributions over X, which is the subspace of vectors with non-negative components $\psi_i(t)$ that sum to 1. The system is subject to some discrete time dynamics given by a stochastic operator $\mathcal{H} : \Lambda^X \to \Lambda^X$. We assume that the dynamics are linear in the following respect:

$$\psi(t+1) = \mathcal{H}\,\psi(t) \quad \text{such that} \quad \psi_i(t+1) = \sum_j \mathcal{H}_{ij}\,\psi_j(t)\,, \tag{1}$$

i.e., a Markov process. The matrix $\mathcal{H}_{ij}$ is also a conditional probability distribution (which we will write as $\mathcal{H}(i|j)$ later) that gives the probability of a state transition from state j to i.

This scenario is compatible with the model of evolutionary adaptation that will be discussed in the next chapter but also generalizes the scenario of neural adaptation discussed in chapter 2. It differs from typical dynamical laws in field theory in two respects: (1) the dynamics is discrete in time, and (2) the field is a real-valued probability field instead of a particle field (which had values in a complex vector space representation of some symmetry; a complex vector bundle). Nonetheless, the following way of capturing the decomposition of dynamics is certainly inspired by particle theory.

Assume the state space X can be decomposed, $X = X^1 \times X^2$. Say, the space X^1 captures one feature of the system and the space X^2 another. The question is whether the evolution of those two features is correlated or independent under the given dynamics. We capture this as follows: If the two features are decorrelated in the initial condition of the system (i.e., in $\psi(t=0)$) then, if the dynamics is decomposed, those two features shall remain decorrelated all through time. In terms of probability distributions, decorrelation means that the distribution is a product of the two marginal distributions:

$$\psi_i = \psi_k^1\,\psi_l^2 \quad \text{for } i = (k, l)\,.$$

Here, $k \in X^1$ and $l \in X^2$ enumerate the states in X^1 and X^2, respectively, and $i = (k, l) \in X = X^1 \times X^2$ is the label of the state in X which corresponds to (k, l). In other words, the field ψ can be represented as a tuple of two fields ψ^1 and ψ^2 (i.e., the direct sum $\psi = \psi^1 \oplus \psi^2$). Now, under which conditions is this decomposition preserved during evolution? As the first case, assume the operator $\mathcal{H}$ is also a product $\mathcal{H}_{ij} = \mathcal{H}_{(k,l)(r,s)} = \mathcal{H}^1_{kr} \mathcal{H}^2_{ls}$. Then,

$$(\mathcal{H}\psi)_{(k,l)} = \sum_{r,s} \mathcal{H}_{(k,l)(r,s)} \, \psi_{(r,s)} = \Big[\sum_k \mathcal{H}^1_{kr} \, \psi^1_r \Big] \Big[\sum_l \mathcal{H}^2_{ls} \, \psi^2_s \Big] \, .$$

Hence, the state remains decomposed. In the second case, when there exist specific $\bar{k}, \bar{l}, \bar{r}, \bar{s}$ such that $\mathcal{H}_{(\bar{k},\bar{l})(\bar{r},\bar{s})} \neq \mathcal{H}^1_{\bar{k}\bar{r}} \mathcal{H}^2_{\bar{l}\bar{s}}$, it is easy find examples for ψ such that decomposition is not preserved (e.g., $\psi_{(r,s)} = \delta_{r\bar{r}} \delta_{s\bar{s}}$ with the so-called delta-distribution δ). Hence, the state remains decomposed if and only if $\mathcal{H}_{(k,l)(r,s)} = \mathcal{H}^1_{kr} \mathcal{H}^2_{ls}$.

There is a more group theoretical way to express this. The evolution equation (1) may be regarded as a group representation, where $\mathcal{H}$ is a group operator that applies linearly on the vector ψ. If $\mathcal{H}$ preserves decomposition and the initial state is decomposed, we may write the vectors ψ as 2-tuples of vectors, $\psi = \binom{\psi^1}{\psi^2}$ and the linear group action becomes decomposed in the sense

$$\mathcal{H}\psi = \begin{pmatrix} \mathcal{H}^1 & 0 \\ 0 & \mathcal{H}^2 \end{pmatrix} \begin{pmatrix} \psi^1 \\ \psi^2 \end{pmatrix} \, . \tag{2}$$

In this representation, the operator $\mathcal{H}$ becomes a block matrix which, in group theory, one would write $\mathcal{H} = \mathcal{H}^1 \oplus \mathcal{H}^2$ and call it a reducible representation of the group action. In elementary particle physics, the reducibility (or rather irreducibility) of representations defines the concept of elementary particles. In a rather analogous way such reducibility of the adaptation operator allows for a general definition of the notion of functional modules in the context of evolutionary and neural adaptation.

Structured adaptation by structured representations. The scenario we just formalized has an important implication for adaptational dynamics. An adaptation step (represented by $\mathcal{H}$) may be triggered by some stochastic event (like for stochastic online learning) or may inherently be stochastic (like for stochastic search and evolution). The decomposability of the operator $\mathcal{H}$ decides whether two features of the system are adapted in correlation or whether they are adapted independently. For example, consider two features of a human, e.g., how s/he rides a bike and how s/he plays the piano. Is it advantageous that both features are adapted in correlation? If a progress in playing the piano entails a corresponding (i.e., correlated) change of how s/he rides the bike, will that be advantageous? Obviously it is important to analyze which neural structures induce an adaptation dynamics $\mathcal{H}$ that can principally be decomposed.

The same fundamental principles of correlated vs. decorrelated adaptation are particularly important in understanding evolutionary adaptation. We will find that for typical computational models of evolutionary adaptation (*evolutionary algorithms*) the "explorative part" of adaptation dynamics is decomposed (selection is generally not). But if we look at nature we find incredibly regular, self-similar structures like tissues, vessels, plants, etc. If every cell of an organism was adapted in decorrelation from other cells it seems impossible that stochastic search would ever find such correlated results. Thus, the key in understanding evolutionary adaptation is to analyze the mechanisms that account for correlations in evolutionary adaptation.

The general approach followed in this thesis to analyze and understand the correlational structure of adaptation dynamics is to investigate the way systems represent their features. This approach considers a non-trivial relation between the ultimate units of adaptation—genes for evolution and synaptic weights for neural systems—and the features of the system. Say, this relation is a mapping $\Phi : G \to X$, where G is the space of genomes (in the simplest case the space of strings over some alphabet) or the space of weight configurations (namely $\mathbb{R}^m$ if weights are considered real-valued). This mapping is an abstraction of any kinds of mechanisms (in nature or in the computer) that determine the relation between the genes and the fitness relevant features of an organism; or between the weights and the functional features of a neural system.

If this mapping is non-trivial, which it certainly is in nature, then it becomes crucial to analyze the correlational structure of adaptation dynamics on both levels, on G and on X. The typical case in evolutionary dynamics is that adaptation is decorrelated on the G-level; genes mutate independently from each other. But the non-trivial relation between G and X may induce highly structured adaptation on the X-level. The reason is that a single gene is responsible for many features of the final organism. If this gene mutates, all these features vary in correlation.

This preliminary discussion, which will be continued in depth throughout the following chapters, shows the importance of how features of a system are represented by the underlying adaptive substrate. In the case of neural systems, the connectional structure together with the operational properties of single neurons functioning determine this relation. In the case of evolutionary adaptation, the genotype-phenotype mapping Φ becomes the central topic and is thoroughly discussed in this thesis.

Chapter 1

Evolutionary adaptation

1.1 Introduction

Evolution can be understood as a search through the space of possible organisms. A computer scientist would also call it the space of possible solution candidates, where to each solution a quality measure is associated. In natural evolution, the notion of fitness is not that simple to define; we will briefly address this issue in section 1.2.3.

Mutation and recombination play a crucial role in this search process. They determine which organisms are possible offspring candidates in the next generation before they are subject to selection; they determine which innovation steps are possible from the parent to the offspring. At first sight it seems that, given the features of a parental organism, the set of possible offspring (we also call it the neighborhood of the parent) is prefixed by the laws of nature or the rules of a genetic algorithm. From a certain point of view, this is not necessarily true. This point of view explicitly distinguishes between the phenotype, i.e., all the functional (selection-relevant) features of an organism, and the genotype, which is the genetic encoding of the phenotype and eventually the underlying substrate subject to evolution. In that language, we can more precisely state the central issue of this chapter:

Given a phenotype, the set of phenotypic neighbors (the set of possible offspring) is *not* necessarily prefixed; instead, the neighborhood of a phenotype depends on its genotype, i.e., the genetic representation of this phenotype.

For instance, we might have two organisms that look exactly the same (in section 1.2.4 we define precisely our use of the word phenotype) but if they are encoded by different genotypes, the probability distribution of their offspring may differ. How might evolution exploit this fact? Suppose evolution found an organism which is functionally quite well adapted but suffers from a lack of innovatability, i.e., all its children are no better than the parent. Now, evolution can change the genetic representation of this organism without changing its functional phenotype. This change of genetic representation, called *neutral mutation*, also changes the organism's offspring neighborhood and in occasion will lead to more innovatability.

On the level of molecular evolution, Schuster (1996) and Fontana & Schuster (1998) analyzed an impressive example for non-fixed phenotypic neighborhoods that depend on the genetic representation. In their studies, the "organisms" are proteins whose functionality is determined by their 3-dimensional molecular shape. This shape is the outcome of a complex folding process and eventually depends on the protein's amino acid sequence—the genotype. The authors showed that the same phenotype (3D shape) can be realized by a large number of different genotypes. Depending on the genotype, the phenotypic neighborhoods change so severely that almost any phenotype becomes possible as an offspring when only a suitable genetic representation is given.

Since the choice of the genetic representation of a phenotype decisively determines the possibilities of phenotypic mutations and innovations, it has been argued that the genetic representations in todays natural organisms are not a mere incident. Instead, they should be the outcome of an adaptive process that optimized these representations with respect to the phenotypic variability and "innovatability" they induce (Wagner & Altenberg 1996). In natural organisms, the way genes represent the functional traits (phenes) of an organism is very complex since the effects of gene expressions highly interact (what is called networks of gene interactions). In fact, the meaning and advantages of this complex relation between genes and phenes are currently widely discussed (Wagner, Booth, & Bagheri-Chaichian 1997; Wagner, Laubichler, & Bagheri-Chaichian 1998; Hansen & Wagner 2001a, 2001b). In trying to understand gene interactions, they are often characterized as advantageous for evolvability (i.e., the chance of exploring new, functionally advantageous phenotypes, Wagner & Altenberg 1996) or as stabilizing mechanisms (e.g., of canalization).

To go beyond arguing for the plausibility of specific genetic representations one should propose a theory on how they *evolve* and what selectional mechanisms guide their evolution. Existing models concentrate on the evolution of gene interactions, e.g., on smooth landscapes (Rice 1998, 2000) or on NK-landscapes that are themselves subject to evolution (Altenberg 1994, 1995). Wagner (1996) also discusses the evolution of directly encoded networks of interacting genes and Wagner & Mezey (2000) propose a multi-linear model of gene interaction.

In the following section, we first formalize the idea of genetic representations and their influence on phenotypic variability. The goal of this formalization is to clarify central issues like the principle meaning of so-called neutral traits for phenotypic evolution while at the same time establishing a relation to mathematical theories on evolutionary algorithms.

The major result is derived on the basis of this formalism. A theorem describes exactly that adaptive process that explains why genetic representations evolve in nature (and should evolve in artificial systems) to encode highly structured and adapted distributions of phenotypic variability. The cornerstones of this theory are the following:

- Neutrality forms the basis of the evolution of gene interactions because it allows that phenotypic variability (formalized by the distribution of phenotypic mutants, the *phenotypic exploration distribution*) is itself variable and adaptable (Riedl 1977; Kimura 1986; Wagner 1996; Rice 1998). Gene interactions are the origin of *structure* in phenotypic variability, where structure means correlations or mutual information between different phenotypic traits within this distribution.

- The evolution of the genetic representations does not imply an evolution of the genotype-phenotype mapping itself. With fixed genotype-phenotype mapping, phenotypically neutral variations may rearrange the genetic system so severely that different gene interactions occur, a different phenotypic variability is induced, and one might even speak of a different genetic encoding. The phenotype is unaffected, but the phenotypic exploration distribution changes.

- The driving force for such rearrangements of the genetic system is the indirect effect of selection on the evolution of exploration distributions. We develop a theoretical model that formulates an evolution equation for exploration distributions and allows to identify the *effective fitness* of exploration distributions guiding their evolution: Exploration distributions are selected with higher probability the better they *match* the fitness distribution over phenotype space; in particular they are selected more likely if they exhibit a correlational structure similar to the correlations between phenotypic traits in selection. Hence, exploration distributions evolve such that dependencies and correlations between phenotypic traits in selection are naturally adopted by the way evolution explores phenotype space.

- Eventually, this allows an information theoretic interpretation of evolutionary dynamics: The information that is given by the selection or non-selection of solutions is implicitly accumulated by evolutionary dynamics and exploited for further search. This information is stored in the way phenotypes are represented. In that way evolution implicitly learns about the problem by adapting its genetic representations accordingly.

The work is related to several current topics of evolutionary computation. First, algorithms have recently been proposed that directly realize an adaptation of exploration distributions by using explicit statistical models for the description of exploration distributions (Estimation-Of-Distributions Algorithms, EDAs, Pelikan, Goldberg, & Lobo 1999; Pelikan, Goldberg, & Cantú-Paz 2000; Baluja & Davies 1997; Mühlenbein, Mahnig, & Rodriguez 1999; Hansen & Ostermeier 2001). More implicitly, also several models of *self-adaptation* are well-established in evolutionary optimization (Rechenberg 1994; Schwefel 1995). We will discuss the relation of such models to our formalism and eventually claim that all these approaches share a basic concept: the incorporation of information, given by the statistics of selection, in future explorations.

Second, the relation between EDAs and crossover in Genetic Algorithms (GAs) is currently discussed in the Evolutionary Computation community (see, e.g., Shapiro 2003, Introduction). Our approach will allow to discuss the implication of crossover in terms of correlated variability in section 1.3. An interesting aspect here is to compare Holland's traditional notion of building blocks of evolution (Holland 1975; Holland 2000) to the notion of functional phenotypic building blocks of an organism that can be motivated in terms of correlated variability (Wagner & Altenberg 1996). Since we will rule out crossover when developing the theory on the evolution of genetic representations in the following section, section 1.3 will also make up for this.

Section 1.4 will discuss some phenomena of natural evolution in terms of the developed theory. The goal is to clarify how the abstract theorems developed relate to nature and increase intuition about them.

Finally, new approaches to evolve complex structures by means of generative (grammar-like) systems (*L-systems*) or so-called *symbiotic composition* have recently proposed (Hornby & Pollack 2001a; Hornby & Pollack 2001b; Watson & Pollack 2002). In the last section of this chapter we present a computational model that will use a similar approach to demonstrate the evolution of genetic representations. The crucial novelty in our model is the introduction of *2nd-type mutations* which we claim is a lack of any other model and allows for neutral rearrangements of the genome and thereby for an evolvability of exploration distributions.

Core parts of this work were first published in (Toussaint 2001; Toussaint 2003b; Toussaint 2003c; Toussaint 2003a). See also (Toussaint & Igel 2002; Igel & Toussaint 2003a) for general discussions on neutrality and self-adaptation, and (Igel & Toussaint 2003b) for discussions and extensions of the No Free Lunch theorem.

1.2   A theory on the evolution of phenotypic variability

1.2.1   A prior in understanding evolution: There is *No Free Lunch* without assuming a constrained problem structure

If you do not presume that the problem has a structure, then any search strategy for good solutions is as good as random search.

Consider a box filled with balls of arbitrary color. Your task is to draw balls, one by one, without replacement from the box and find a red ball as quickly as possible. The point is that, if there is no constraint on how the balls (respectively colors) are distributed in the box (e.g., black ones to the left and red ones to the right, etc.), then there exists no strategy for picking balls that is superior to another. Here, no constraint on the distribution of balls means not even the constraint that there *exists* a detectable kind of structure in the distribution of balls. Otherwise there might be a strategy to learn about this structure and thereafter exploit this knowledge.

This fact is often discussed in the context of evolutionary algorithms because they have been said to be "general purpose search algorithms". The example above though illustrates that this cannot be true without making any assumption about the problem, i.e., about the distribution of solutions. The theorem is known as "No Free Lunch" and was introduced by Wolpert & Macready (1995, 1997). Stated more rigorously, though still in words, it says

Theorem 1.2.1 (No Free Lunch (Wolpert & Macready 1995)). *Consider the problem of drawing search points from a finite set. A quality measure is associated with each point. Further, the distribution of quality measures over the points is completely unconstrained and unknown: Every permutation of quality measures is equally likely. A search strategy (algorithm) may be evaluated after drawing n points by looking at its history of quality measures it drew.*

Now, on average, that is when averaged over all possible problems (which essentially corresponds to averaging over all permutations of quality measures), the average history of quality measures is the same for any two search strategies. Thus, any evaluation of two search strategies must be equal for any n.

The theorem was formalized and generalized in many ways. Schuhmacher, Vose, & Whitley (2001) generalized it to be valid for any subsets of problems that are closed under permutation. Igel & Toussaint (2003b) derived sufficient and necessary conditions for No Free Lunch when the probability of permutations is not equally but arbitrarily distributed. In that paper we also argued that generally the necessary conditions for No Free Lunch are hardly ever fulfilled.

We do not want to present the details of that research at this place. The reason the No Free Lunch theorem is mentioned at the beginning of a chapter on evolution is that it clearly states what it means to make assumptions about the problem, namely, to make assumptions about the distribution or the existence of a structured distribution of quality measures over all search points. Only when making this assumption, evolutionary search has a chance to learn about and exploit this structure of the distribution of good solutions and efficiently explore the space of solutions.

Evolutionary search explores the search space by means of mutation and recombination, which defines a *variational topology* on the search space. Intuitively, this kind of search will be efficient if the fitness function is not completely random but more or less continuous with respect to this variational topology such that continuous evolutionary progress becomes possible. The continuity of fitness on the variational topology has been captured by the definition of *strong causality* (Rechenberg 1994; Sendhoff, Kreutz, & von Seelen 1997), a concept which in turn may be used for the design of mutation operators that allow for more continuous progress.

However, in natural evolution mutation operators are not designed by some intelligence. A central question arises: What does it mean to "learn" about the problem structure and exploit it? How in principle can evolution realize this? The answer we will give is that the implicit process of the evolution of genetic representations allows for the self-adaptation of the "search strategy" (i.e., the phenotypic variability induced by mutation and recombination). To some degree, this process has been overlooked in the context of evolutionary algorithms because complex, non-trivial (to be rigorously defined later) genetic representations (genotype-phenotype mappings) have been neglected by theoreticians. This chapter tries to fill this gap and propose a theoretical framework for evolution in the case of complex genotype-phenotype mappings focusing at the evolution of phenotypic variability. The next section lays the first cornerstone by clarifying what it means to learn about a problem structure.

1.2.2   How to assume and learn about a problem structure: A generic heuristic to adapt the search distribution

The structure of a problem is described by the structure of the distribution of good solutions—learning the structure of a problem means to adapt the search distribution toward the distribution of good solutions.

If we assume that there exists an identifiable structure in the distribution of good solutions to a problem, then there is a chance to learn about this structure and to make search directed and more efficient. The framework of such an approach is formalized, e.g., by Zhigljavsky (1991): Let P be the search space and $f : P \to \mathbb{R}^+$ the quality measure defined for each search point. By Λ^P we denote the space of distributions[1] over P, i.e., the space of all functions $P \to [0, 1]$ that integrate to 1. In an algorithmic fashion, *global random search* is described by:

(1) Choose a probability distribution $q^{(t)} \in \Lambda^P$ on P.

(2) Sample λ points $x_1^{(t)}, \dots, x_\lambda^{(t)}$ from the distribution $q^{(t)}$ and evaluate the quality f (perhaps with random noise) at these points.

(3) According to a fixed heuristic rule (which depends on the algorithm) construct a new probability distribution $q^{(t+1)}$ depending on the evaluations.

(4) Check for some appropriate stopping condition; if the algorithm has not terminated, substitute $t := t + 1$ and return to Step 2.

Let us rewrite this definition in a way that is more consistent with the conventions we use throughout this chapter, leading to a discrete-time stochastic dynamic equation of some parameters $y^{(t)}$ of the search distribution $q^{(t)} = \Phi y^{(t)}$. So we assume that $\Phi : Y \to \Lambda^P$ is a parameterization of distributions, i.e., introduces coordinates on the space of distributions, where Y is the parameter space.

Next, we write the sampling of λ points in the second step as a stochastic operator

---

[1] Throughout this work, we will use the word "distribution" for both, probability densities over continuous spaces as well as probability distributions over finite spaces. We do not need to make an assumption on whether the search space P is discrete or continuous. In the continuous case, when writing a summation over P it should be understood as an integration.

$S^\lambda : \Lambda^P \to \Lambda^P$ mapping any distribution q to the *finite distribution*

$$S^\lambda q := \frac{1}{\lambda} \sum_{i=1}^{\lambda} \delta_{x_i} , \quad x_i \text{ are } \lambda \text{ independently drawn samples of } q ,$$

where $\delta_{x_i} \in \Lambda^P$ is the delta-distribution at the sampling point x_i. Note that $\lim_{\lambda \to \infty} S^\lambda = \text{id}$, i.e., the identity. We represent also finite populations as a distribution $p \in \Lambda^P$, namely, if the population is given as a finite multiset $A = \{x_1, .., x_\mu\}$ we (isomorphically) represent it as the finite distribution given by $p = \frac{1}{\mu} \sum_{i=1}^{\mu} \delta_{x_i}$, i.e., $p(x) = \frac{|A \cap \{x\}|}{|A|} = \frac{\text{multiplicity of } x \text{ in } A}{|A|}$.

The evaluations of the quality f mentioned in the second step of global random search are written as an operator $\mathcal{F} : \Lambda^P \to \Lambda^P$ that maps any distribution q to the linearly rescaled distribution

$$(\mathcal{F} q)(x) = \frac{f(x)}{\bar{f}} q(x) , \quad \text{with } \bar{f} := \sum_{x \in P} f(x) q(x) . \tag{1.1}$$

Putting this together, the above algorithm may be defined as follows.

Definition 1.2.1 (Heuristic search). Given a search space P and a quality measure $f : P \to \mathbb{R}^+$ defined for each search point, global random search may be described by the stochastic iteration equation

$$y^{(t+1)} = \mathcal{H}(\mathcal{F} \tilde{q}^{(t)}, \tilde{q}^{(t)}, y^{(t)}) , \quad \text{where } \tilde{q}^{(t)} = S^\lambda q^{(t)} , \quad q^{(t)} = \Phi y^{(t)} . \tag{1.2}$$

Herein, $\mathcal{H}$ represents the heuristic rule that determines the new distribution parameters $y^{(t+1)}$ depending on: (1) the old distribution parameters $y^{(t)}$, (2) the set of λ sample points given uniquely by the sampled finite distribution $\tilde{q}^{(t)}$, and (3) by the evaluations $\mathcal{F} \tilde{q}^{(t)}$ which encodes all the information on the qualities of the samples. (This definition is identical to the algorithmic definition above under two constraints: The heuristic rule must be scale invariant (i.e., independent of rescaling evaluations by $1/\bar{f}$) and evaluation need to be non-negative (if negative, $\mathcal{F}$ would not be a distribution operator).

One might complain that this notation did not simplify anything. This is true for the general heuristic search. We will though find below that, when written in this formalism, there exist a generic choice for the heuristic rule $\mathcal{H}$ which simplifies equation (1.2) severely.

The core ingredient of this search scheme is the search distribution $q^{(t)} = \Phi y^{(t)}$, that we also call *exploration distribution*, on the search space. This distribution is used to represent the knowledge or assumptions made about the true distribution of good solution. The heuristic rule is supposed to adapt these assumptions according to the experienced evaluations.

Many search algorithms, essentially also evolutionary algorithms (Toussaint & Igel 2002), belong to the class of global random search algorithms but differ in their specific heuristic rules and in the way they parameterize the search distribution, i.e., they differ in Φ. In the following we want to define a *generic* heuristic rule that captures a very general point of how the adaptation of the search distribution can in principle exploit the information given by the evaluations. The central question is, how can "the information given by the evaluations" be captured formally and how can this information be accumulated and exploited in the search strategy $q^{(t)}$?

We capture the "information given by the evaluations" as the *difference* between the search distribution $q^{(t)}$ and the distribution $\mathcal{F} \mathcal{S}^\lambda q^{(t)}$ after sampling and evaluation. The difference can by quantified by an information theoretic measure, the *relative entropy* or *Kullback-Leibler divergence* (Kullback & Leibler 1951; Cover & Thomas 1991) which is defined as follows: Given two distributions $q \in \Lambda^P$ and $p \in \Lambda^P$ the Kullback-Leibler divergence $D(p : q)$ reads

$$D(p : q) := \sum_x p(x) \ln \frac{p(x)}{q(x)} . \tag{1.3}$$

It is a measure for the loss of information (or gain of entropy) when a *true* distribution p is represented (approximated) by a model distributions q. For example, when $p(x, y)$ is approximated by $p(x) p(y)$ one looses information on the mutual dependence between x and y. Accordingly, the relative entropy $D(p(x, y) : p(x) p(y))$ is equal to the mutual information between x and y. Generally, when *knowing* the real distribution p one needs on average H_p (entropy of p) bits to describe a random sample. If, however, we know only an approximate model q we would need $H_p + D(p : q)$ bits to describe a random sample of p. The loss of knowledge about the true distribution induces an increase of entropy and thereby an increase of description length for random samples.

The Kullback-Leibler divergence between the evaluated distribution $\mathcal{F} \mathcal{S}^\lambda q^{(t)}$ and another distribution q' thus measures whether solutions with high quality are also likely in q', but also whether correlations between solution features or other kinds of structure in the evaluated distribution are also present in q'. Concerning the adaptation of the search distribution $q^{(t)}$, information can be accumulated and exploited by adapting $q^{(t)}$ such that solutions with high evaluation become more probable than others. For instance, correlations between solution features found in $\mathcal{F}\,\mathcal{S}^\lambda\,q^{(t)}$ can be incorporated in new explorations $q^{(t+1)}$. This may be realized by adapting $q^{(t)}$ such that the difference between the just experienced evaluation distribution $\mathcal{F}\,\mathcal{S}^\lambda\,q^{(t)}$ and the forthcoming search distribution $q^{(t+1)}$ becomes small. This is what we call the generic heuristic rule:

Definition 1.2.2 (Generic heuristic search (GHS)). Given a search space P, a parameterization $\Phi : Y \to \Lambda^P$ of search distributions over P, and an evaluation operator $\mathcal{F} : \Lambda^P \to \Lambda^P$, then *generic heuristic search* is given by

$$y^{(t+1)} = \operatorname*{argmin}_{y \in Y}\ D\big(\mathcal{F}\,\mathcal{S}^\lambda q^{(t)} : \Phi y\big) . \tag{1.4}$$

In case Y is continuous, we also define a *continuous generic adaptation* by:

$$y^{(t+1)} = (1 - \alpha)\, y^{(t)} + \alpha \operatorname*{argmin}_{y \in Y}\ D\big(\mathcal{F}\,\mathcal{S}^\lambda q^{(t)} : \Phi y\big) , \tag{1.5}$$

with *adaptation rate* $\alpha \in [0, 1]$.

The role of the parameterization Φ will be discussed in more detail in the following sections. If the parameterization is such that all distributions are representable (Φ is surjective), then equation (1.4) reduces to $q^{(t+1)} = \mathcal{F}\,\mathcal{S}^\lambda\,q^{(t)}$. This will though never be the case in the remainder of this work.

One can grasp equation (1.4) intuitively is as follows. The new state $q^{(t+1)}$ tries to approximate the evaluated old state $\mathcal{F}\,\mathcal{S}^\lambda q^{(t)}$ (the donkey alike, chasing the carrot on a stick, where $\mathcal{F}\,\mathcal{S}^\lambda$ is the stick). The driving force of this recursion is the Kullback-Leibler divergence between $\mathcal{F}\,\mathcal{S}^\lambda q^{(t)}$ and $q^{(t)}$ (the distance between carrot and donkey), which is the new information. If this divergence is minimal, dynamics reach a fixed point $q^{(t+1)} = q^{(t)}$. The fixed point may be considered a "pseudo-goal" of search—actually a quasi-species (Eigen, McCaskill, & Schuster 1989). In the context of evolution we will deal with search distributions of non-vanishing entropy (excluding the δ-peak as trivial fixed point) and focus on the discussion of their *structure*, i.e., the mutual correlations and dependencies between solution features within the distributions, which are also captured by the Kullback-Leibler divergence and, according to (1.4), subject to adaptation.

Given a problem at hand, generic heuristic search is uniquely defined by the choice of Φ. There exists at least one algorithm in the literature that exactly realizes GHS: the Bayesian Optimization Algorithm (BOA, Pelikan, Goldberg, & Cantú-Paz 2000), which uses general Bayes networks to parameterize the search distribution. There are many algorithms that realize GHS *almost* exactly: Population-Based Incremental Learning (PBIL, Baluja 1994) uses a simple product distribution as the search distribution, Baluja & Davies (1997) use a dependency tree to parameterize the search distribution, and Mühlenbein, Mahnig, & Rodriguez (1999) use a factorized distribution model. The difference between these algorithms and GHS are subtleties in the update rule (typically a mixture between continuous generic adaptation for some parameters and discrete generic adaptation for others; or other, model specific or approximate distance measures rather than the exact Kullback-Leibler divergence are used). In the next section we will also show how standard evolutionary algorithms can be understood as an approximation of GHS.

Hence, our definition of GHS gives an abstract way of understanding a whole class of adaptation processes. But above that, the reason we defined GHS is that it pinpoints what exactly we mean by "learning about a problem structure": namely incorporating the information given by evaluations (in particular correlations) in forthcoming search. GHS will become the underlying metaphor in the following discussions of the evolution of phenotypic variability.

1.2.3 Evolutionary processes: A population, mutations, and recombination to represent the search distribution

Evolution can be described as a stochastic dynamic process on the genotype space, comparable to generic heuristic search, where the search distribution is parameterized by a finite parent population via mutation and recombination.

Let us turn from heuristic search specifically to evolutionary systems. How is the search distribution parameterized and adapted in evolutionary systems? We introduce evolution as a stochastic dynamic process that can be embedded in the heuristic search scheme:

Definition 1.2.3 (Evolutionary processes). The free variables of the process (which we denoted by $y^{(t)}$ above) are a distribution $p^{(t)} \in \Lambda^G$ of parents. In the finite population case, $p^{(t)}$ is a finite distribution. The space G is called genotype space and we write the general stochastic process that describes evolution as

$$p^{(t+1)} = \mathcal{S}^{\mu} \, \mathcal{F} \, \mathcal{S}^{\lambda} \, \mathcal{M} \, p^{(t)} \, . \tag{1.6}$$

We will discuss the concepts "genotype" and "phenotype" in more detail in the next section. Roughly, what this equation describes is a parent population $p^{(t)}$ that, via mutation and recombination, defines a probability distribution $\mathcal{M} \, p^{(t)}$ of offspring. From this distribution a population of λ offspring is sampled via $\mathcal{S}^{\lambda}$ and evaluated by the fitness operator $\mathcal{F}$. Proportional to the fitness, a population of μ parents is selected by sampling the fitness distribution over the offspring population. We discuss each step of the process in more detail in the following.

Mutation and recombination as the parameterization Φ of the search distribution. The search or *exploration* on the genotype space is governed by the operator $\mathcal{M} : \Lambda^{G} \to \Lambda^{G}$ (called *mixing*) which accounts for mutation and recombination. Namely, the *offspring distribution* $q^{(t)} = \mathcal{M} \, p^{(t)}$ describes the probability of an offspring given the parent population $p^{(t)}$ and corresponds to the search distribution in the heuristic search scheme. In summary, the free variable of evolutionary dynamics is the parent population $p^{(t)} \in \Lambda^{P}$ such that the search distribution is the mixing of the parent distribution, $q^{(t)} = \Phi y^{(t)} = \mathcal{M} \, p^{(t)}$. This points out that, *the mixing operator is the parameterization Φ of the search distribution for evolutionary processes.*

This also implies that the space $\Phi(Y) = \mathcal{M}(\Lambda^{G})$ of representable search distributions is limited to those distributions that are an outcome of the mixing operator,

$$\{\mathcal{M} \, p \mid p \in \Lambda^{G}\} \, .$$

E.g., when neglecting crossover, each individual in $p^{(t)}$ basically contributes an additive component to the search distribution, which is given by the mutation distribution of this individual. This is very similar to how continuous distributions are often encoded for numeric approximations: as a finite sum of kernels, typically Gaussian kernels. Special about this way of parameterizing the search distribution is that it naturally allows to represent multi-modal distributions—in contrast to other search schemes like the Evolution Strategies (Rechenberg 1994; Schwefel 1995). Multi-modality means that search can be performed parallel in different regions.

Offspring sampling.   As for heuristic search, only a finite number of offspring are actually realized: The stochastic sample operator $\mathcal{S}^\lambda$ maps the offspring probability distribution $q^{(t)}$ to a finite population $\tilde{q}^{(t)} = \mathcal{S}^\lambda q^{(t)}$ of λ individuals.

Fitness.   The operator $\mathcal{F} : \Lambda^G \to \Lambda^G$ was defined in equation (1.1, page 25) and associates a selection probability $f^{(t)}(g)/\bar{f}^{(t)}$ to each member g of the offspring population. We call this operator the *fitness operator* and define

Definition 1.2.4 (Fitness). Given an evolutionary process in form of equation (1.6), we call the function $f^{(t)}$, as given in the definition (1.1) of the operator $\mathcal{F}$, the *fitness function*. In other words, *given* an evolutionary process we define an individual's fitness as its current selection probability. Thus, fitness may generally depend on time and on the rest of the population (in the case of so-called "tournament" or "ranking" selection schemes).

Since we *derive* the notion "fitness" from a presumed existing evolutionary process, our usage of the word fitness is in agreement to how most biologists would use it—"survival of the fittest" becomes a tautology. In contrast, in the context of evolutionary computation, the notion fitness is usually not derived from an existing evolutionary process to describe the process, but an ad hoc fitness function is assumed and enters the *definition* of the algorithm by the design of the selection scheme.

Applying $\mathcal{F}$ on the offspring population $\tilde{q}^{(t)}$ one gets a distribution over the offspring population that we call *fitness distribution*.

Selection as (an approximate generic) heuristic rule $\mathcal{H}$.   Finally, the sample operator $\mathcal{S}^\mu$ performs selection by mapping the fitness distribution $\mathcal{F}\tilde{q}^{(t)}$ to the new parent population $p^{(t+1)}$. In evolution, selection $\mathcal{S}^\mu$ plays the role of the heuristic rule $\mathcal{H}$: Instead of a general adaptation scheme that exploits the information given in the fitness distribution, this distribution is directly sampled to generate the new parent population. Sampling the fitness distribution may be interpreted as finite approximation of the fitness distribution, and, as we discussed above, this finite approximation (the parent population) is in turn similar to a finite approximate encoding of the new search distribution. Thus, the evolution process can be regarded as realizing an approximate generic heuristic search, i.e., the generic way of accumulating information during search by adapting the search distribution toward the experienced fitness distribution.

Let us analyze this more precisely. In the original notation, the generic heuristic rule reads $y^{(t+1)} = \underset{y \in Y}{\arg\min} \; D\big(\mathcal{F}\,\mathcal{S}^\lambda\,\Phi y^{(t)} : \Phi y\big)$. This translates to the case of evolutionary dynamics if we replace the search parameters $y^{(t)}$ by the population $p^{(t)}$, the parameter space Y by the space of μ-finite populations $\Lambda^{\mu,G} \subset \Lambda^G$, and the distribution model Φ by $\mathcal{M}$. Hence the question is, under which conditions it is true that

$$\underset{p' \in \Lambda^{\mu,G}}{\arg\min} \; D\big(\mathcal{F}\,\mathcal{S}^\lambda\,\mathcal{M}p^{(t)} : \mathcal{M}p'\big) \; = \; \mathcal{S}^\mu\,\mathcal{F}\,\mathcal{S}^\lambda\,\mathcal{M}p^{(t)} \; .$$

If the equation is fulfilled, then evolutionary dynamics and generic heuristic search coincide. If we impose no conditions on the population $p^{(t)}$ and the selection operator $\mathcal{F}$ this simplifies to the question, for which operators $\mathcal{M}$ it is true that

$$\forall\, s \in \Lambda^{\lambda,G} \; : \quad \underset{p' \in \Lambda^{\mu,G}}{\arg\min} \; D\big(s : \mathcal{M}p'\big) \; = \; \mathcal{S}^\mu\, s \; .$$

We cannot solve this problem in all generality here. Intuition tells that, if $\mathcal{M}$ is a "reasonable" mutation operator that corresponds to small mutations of individuals and only smoothes the population distribution, this equation is approximately true. But we can make this more explicit for a widely assumed component-wise mutation operator (which we will investigate in depth in section 1.3). Let the probability of non-mutation be constant, $\mathcal{M}(x|x) = \alpha$, and let $\mathcal{M}(x|y)|_{x \neq y} = \epsilon$. Then we find

$$\mathcal{M}p(x) = \sum_y \mathcal{M}(x|y)\,p(y) = \sum_{y \neq x} \epsilon\,p(y) + \alpha\,p(x) = \epsilon(1 - p(x)) + \alpha\,p(x)$$

$$D\big(s : \mathcal{M}p\big) = \sum_x s(x) \ln \frac{s(x)}{\mathcal{M}p(x)} = \sum_x s(x) \ln \frac{s(x)}{\epsilon(1 - p(x)) + \alpha\,p(x)}$$

$$\leq \sum_x s(x) \ln \frac{s(x)}{\alpha\,p(x)} = D\big(s : p\big) + \ln \frac{1}{\alpha}$$

The smaller epsilon is, the tighter this upper bound. In this sense we may conclude that minimizing $D\big(s : \mathcal{M}p\big)$ approximately equals the problem of minimizing $D\big(s : p\big)$. But since p needs to be finite, $D\big(s : p\big)$ is minimized by a finite approximation of s, i.e., it is minimized by $\mathcal{S}^\mu s$ (where $\mathcal{S}^\mu$ should minimize the sampling error as it is the case, e.g., for the stochastic universal sampling).

Hence, for this special mutation operator we may assume that

$$\forall s \in \Lambda^{\lambda,G} \quad : \quad \operatorname*{argmin}_{p' \in \Lambda^{\mu,G}} D\big(s : \mathfrak{M}p'\big) \approx \mathcal{S}^{\mu}\, s$$

and ordinary selection is indeed an approximation of the generic heuristic.

The genotype space G.   Finally, we defined evolution as a process on a geno-type space G instead of some search space P. The relation between these two spaces will be the topic of the next section.

1.2.4   The genotype and phenotype: Reparameterizing the search space

The mapping from genotype to phenotype is the key to understand com-plex phenotypic variability and evolution's capability to adapt the search distribution on the phenotype space.

We defined evolution as a process on a genotype space G. We argue now, that not G but the so-called phenotype space P should be considered as the actual search space of evolution. A genotype-phenotype mapping $\phi : G \to P$ with the defining property that fitness depends only on the phenotype, $f^{(t)}(g) = \tilde{f}^{(t)}(\phi(g))$, connects both spaces. The implication of this "reparameterization" of the search space will be a topic in the remainder of this chapter. To clarify these concepts, we review the general definitions of phenotype and genotype:

Definition 1.2.5 (Complete and partial phenotype, phenotype space).

(1) In nature, where we presume to know the definition of an individual, the *complete phenotype* of an individual is the collection (tuple) of all (measurable) characters of this individual. Hence, an individual's selection-relevant features as well as the DNA are part of the complete phenotype.

(2) A *partial phenotype* is a lower dimensional projection of the complete pheno-type. In natural evolution one refers to all kinds of partial phenotypes (pheno-typic traits; phenes); in evolutionary computation one typically refers to exactly that partial phenotype on which the ad hoc given fitness function is defined.

(3) We will use the word *phenotype* for the minimal partial phenotype that com-prises all selection-relevant phenotypic traits, such that fitness depends only on

this phenotype. The *phenotype space* P is the space of these phenotypes. In the language of heuristic search algorithms, P comprises all feasible "candidate solution to a problem" such that it is identical to the search space as we defined it for heuristic search.

The genotype has been implicitly defined as the free variables of the evolutionary process. In biology, the genotype is analogously defined as all inherited traits:

Definition 1.2.6 (Genotype). An individual's *genotype* is the collection (tuple) of those characters that are inherited, i.e., formally, the variables that specify the offspring distribution and are subject to the evolution equation (1.6, page 29). Interesting is that in natural evolution, according to this definition, one should, for instance, also regard the ovum also as part of the genotype—which is hardly ever done, but as we will do in the computational model in section 1.5.

The *genotype-phenotype mapping (GP-map)* $\phi : G \to P$ is actually a simplification of the concepts of ontogenesis and innateness because it describes how the genotype determines the phenotype. This mapping may be understood as a transformation or reparameterization of the search space. However, the implications of a GP-map, in particular if it is non-injective, are severe and will be the main subject of this chapter. The non-injectiveness of ϕ may be described in terms of an equivalence relation:

Definition 1.2.7 (Phenotype equivalence, neutral sets). We define two genotypes g_1 and g_2 equivalent iff they have the same phenotype,

$$g_1 \equiv g_2 \iff \phi(g_1) = \phi(g_2) .$$

The set of equivalence classes $G/\equiv$ is one-to-one with the set $P = \phi(G) = \{\phi(g)|g \in G\}$ of phenotypes. Thus, we use the phenotype $x \in P$ to indicate the equivalence class

$$[x] := \phi^{-1}(x) = \{g \in G \,|\, \phi(g) = x\} ,$$

which we also call *neutral set* of $x \in P$ or *phenotypic class*.

The GP-map $\phi : G \to P$ also induces a lift of a genotype distribution onto the phenotype space:

Definition 1.2.8 (Phenotype projection, distribution equivalence). Given a genotype distribution $p \in \Lambda^G$ we define the projection Ξ to a phenotype distribution $\Xi p \in \Lambda^P$ by

$$\Xi p(x) = \sum_{g \in [x]} p(g) \ .$$

By this projection, the equivalence relation carries over to distributions in Λ^G: Two distributions p_1, p_2 are called equivalent iff they induce the same distributions over the phenotype space,

$$p_1 \widehat{\cong} p_2 \iff \Xi p_1 = \Xi p_2 \ .$$

Again, the quotient space $\Lambda^G / \widehat{\cong}$ of equivalence classes in Λ^G is one-to-one with the space Λ^P of distributions over the phenotype space.

For example, the *neutral degree* may be defined by this projection:

Definition 1.2.9 (Neutral degree). The *neutral degree* $n(g)$ of a genotype $g \in G$ is the probability that mutations do not change the phenotype, i.e.,

$$n(g) = [\Xi \mathcal{M}(\cdot | g)] \left(\phi(g) \right) = \sum_{g' \in [\phi(g)]} \mathcal{M}(g' | g) \ .$$

1.2.5   The topology of search: Genotype versus phenotype variability

A simple, by mutation induced genotypic variability may, via a non-injective genotype-phenotype mapping, lead to arbitrarily complex phenotypic variability.

The sampling, fitness, and selection operators in the evolution equation (1.6, page 29) are *non-explorative*: If $p(g) = 0$ for some genotype g, it follows that $(\mathcal{S}^\lambda q)(g) = (\mathcal{F} \mathcal{S}^\lambda q)(g) = 0$. This means that if a genotype has zero probability in a genotype population (e.g., in a finite population), it has still has zero probability after sampling, evaluation, or selection; these operators do not generate *new* genotypes. Search without an explorative component would be trapped in a small region of the search space.

In the evolution equation it is the mutation operator $\mathcal{M}$ that represents *exploration*. Let us neglect crossover here and postpone the discussion of the special features of exploration with crossover until section 1.3. We may formalize mutations as a conditional probability distribution $\mathcal{M}(g'|g)$, giving the probability of a mutation from one genotype $g \in G$ to another $g' \in G$. Then, the *genotypic* offspring distribution is given by $(\mathcal{M}\,p^{(t)})(g') = \sum_g \mathcal{M}(g'|g)\,p^{(t)}(g)$. The conditional probability $\mathcal{M}(g'|g)$ describes how the genotype space G can be explored and actually there exists no (mathematical) structure on G other than the conditional probability, i.e., G is but a set of possible genotypes without any a priori metric or topological structure. The mutation probability $\mathcal{M}$ though induces a *variational structure* on G which is often associated to a variational topology:

Definition 1.2.10 (Genotypic variational topology). Given the conditional mutation probability $\mathcal{M}(\cdot|\cdot)$ on the genotype space, we may define g' a neighbor of g if $\mathcal{M}(g'|g)$ is greater than some limit ϵ. We call such a topology on the genotype space a *genotypic variational topology*, see (Stadler, Stadler, Wagner, & Fontana 2001).

It is intuitive to discuss fundamental features of evolutionary transitions in terms of this variational topology (Schuster 1996; Fontana & Schuster 1998; Reidys & Stadler 2002). What is most interesting for us is the particular role of the GP-map in this context because it induces a variational topology on the phenotype space depending on the topology of the genotype space. The role of the projection Ξ becomes more evident: While $\mathcal{M}(\cdot|g)$ describes the variational structure on G, its projection $\Xi\mathcal{M}(\cdot|g)$ describes the variational structure on P. The projection Ξ may be interpreted as a *lift* of variational structure from the genotype space onto the phenotype space. This is in analogy to the introduction of local coordinates on a manifold (cp. phenotype space) by a local map from a base space of variables (cp. genes). While on the base space usually a Cartesian metric (cp. non-correlating gene variability) is assumed, the induced metric on the local coordinates is arbitrary (cp. correlations between phenotypic traits in phenotypic variability, cf. also to the epistasis matrix of Rice 1998).

There is, however, a crucial difference: The map ϕ need not be one-to-one. If ϕ is non-injective, there exist different genotypes g_i that map to the same phenotype. And thus there exist different neighborhoods U_{g_i} that map to *different* neighborhoods of the *same* phenotype. Hence, the variational topology on phenotype space is generally not fixed but variable and depends on the genotypic representation g_i that induces the topology!

1.2.6   Commutativity and neutrality: When does phenotype evolution depend on neutral traits?

Neutral traits (of which strategy parameters are a special case) have an impact on phenotype evolution if and only if they influence mutation probabilities and thus encode for different exploration distributions.

Based on this formalism we investigate the following question: When $\mathcal{G}$ describes the evolutionary process on the genotype level and if we observe only the projected process on the phenotype level, can we model the phenotypic evolution process without reference to the genotype level? This means, is the genotype level irrelevant for understanding the phenotypic process? Formally, this amounts to whether $\mathcal{G}$ is compatible with the equivalence relation $\equiv$ or not: An operator $\mathcal{G} : \Lambda^G \to \Lambda^G$ is compatible with an equivalence relation $\equiv$ iff

$$p_1 \widehat{\equiv} p_2 \implies \mathcal{G}(p_1) \widehat{\equiv} \mathcal{G}(p_2) \,, \tag{1.7}$$

which is true iff there exists an operator $\tilde{\mathcal{G}}$ such that the diagram

$$
\begin{array}{ccc}
\Lambda^G & \xrightarrow{\ \mathcal{G}\ } & \Lambda^G \\
\Xi\downarrow & & \downarrow\Xi \\
\Lambda^P & \xrightarrow{\ \tilde{\mathcal{G}}\ } & \Lambda^P
\end{array}
\tag{1.8}
$$

commutes, i.e., $\Xi \circ \mathcal{G} = \tilde{\mathcal{G}} \circ \Xi$. This means that, in the case of compatibility, one can define a process $\tilde{\mathcal{G}} : \Lambda^P \to \Lambda^P$ solely on the phenotypic level that equals the projected original process $\mathcal{G}$. In this case, $\tilde{\mathcal{G}}$ represents the phenotypic version of evolutionary dynamics and the population Ξp of phenotypes evolves *independent* of the neutral traits within the genotype population p. Accordingly, we make the following definition:

Definition 1.2.11 (Trivial neutrality, trivial genotype-phenotype mapping (Toussaint 2003b)). Given a genotype space G, an evolutionary process $\mathcal{G} : \Lambda^G \to \Lambda^G$, and a genotype-phenotype mapping $\phi : G \to P$, we define the GP-map *trivial* iff phenotype equivalence $\equiv$ commutes with the evolutionary process $\mathcal{G}$. In that case we also speak of *trivial neutrality*.

The meaning of this definition should become clear from the definition of compatibility: In the case of trivial neutrality, the evolution of phenotypes can be completely understood (i.e., modeled) without referring at all to genotypes, in particular, neutral traits are completely irrelevant for the evolution of phenotypes. We can derive exact conditions for the case of trivial neutrality:

Theorem 1.2.2 ((Toussaint 2003b)). *Let the evolution operator* $\mathcal{G} = \mathcal{F}\mathcal{M}$ *be composed of selection and mutation only (no crossover), and let* $\mathcal{M}$ *be given by the conditional probability* $\mathcal{M}(g'|g)$ *of mutating a genotype* g *into* g'. *Then, neutrality is trivial iff*

$$\forall x \in P \ : \ g_1, g_2 \in [x] \Rightarrow \Xi\mathcal{M}(\cdot|g_1) = \Xi\mathcal{M}(\cdot|g_2) \ .$$

In other words, neutrality is non-trivial if and only if there exists at least one neutral set $[x]$ *such that the projected exploration distribution* $\Xi\mathcal{M}(\cdot|g) \in \Lambda^P$ *is non-constant over this neutral set (i.e., differs for different* $g \in [x]$*).*[2]

Proof. Since selection depends only on phenotypes it is obvious that the selection operator $\mathcal{F}$ commutes with phenotype projection. The composition of two compatible operators is also compatible (Vose 1999, Theorem 17.4). Hence, we need to focus only on the mutation operator $\mathcal{M}$:

Let us consider the mutational process given by

$$p^{(t+1)}(g') = \sum_g \mathcal{M}(g'|g)\, p^{(t)}(g)$$

Following definition (1.7) of compatibility we investigate what happens under projection Ξ:

$$\Xi p^{(t+1)}(x') = \sum_{g' \in [x']} \sum_g \mathcal{M}(g'|g)\, p^{(t)}(g) = \sum_g \sum_{g' \in [x']} \mathcal{M}(g'|g)\, p^{(t)}(g)$$
$$= \sum_g \Xi\mathcal{M}(x'|g)\, p^{(t)}(g) \ .$$

We distinguish two cases:

First case: For all neutral sets $[x]$, let $\Xi\mathcal{M}(\cdot|g)$ be constant over the neutral set, i.e., independent of $g \in [x]$, and we can write $\Xi\mathcal{M}(\cdot|g) = \Xi\mathcal{M}(\cdot|x)$. It follows:

$$\Xi p^{(t+1)}(x') = \sum_x \sum_{g \in [x]} \Xi\mathcal{M}(x'|g)\, p^{(t)}(g) = \sum_x \Xi\mathcal{M}(x'|x) \sum_{g \in [x]} p^{(t)}(g)$$

---

[2]We use the "·" notation as a "wild card" for a function argument. E.g., given a function $f \ : \ \mathbb{R}^2 \to \mathbb{R} \ : \ (x,y) \mapsto f(x,y)$, if we want to fix y and consider the function on that hyperplane, we write $f(\cdot,y) : \mathbb{R} \to \mathbb{R} : x \mapsto f(x,y)$. Accordingly, for a conditional probability we write $\mathcal{M} \ : \ G \to \Lambda^G \ : \ g \mapsto \mathcal{M}(\cdot|g) \in \Lambda^G$.

$$= \sum_x \Xi\mathcal{M}(x'|x)\,\Xi p^{(t)}(x)\;.$$

Hence, $\mathcal{M}$ is compatible with phenotype equivalence; the diagram (1.8) commutes with the "coarse-grained" mutation operator $\widetilde{\mathcal{M}}$ given by $\Xi\mathcal{M}(x'|x)$.

Second case: Let there exist at least one neutral set $[x]$ with different genotypes $g_1, g_2 \in [x]$ such that the corresponding projected exploration distributions are not equal, $\Xi\mathcal{M}(\cdot|g_1) \neq \Xi\mathcal{M}(\cdot|g_1)$. Further, consider the two single genotype populations $p_1^{(t)}(g) = \delta_{g,g_1}$ and $p_2^{(t)}(g) = \delta_{g,g_2}$, which are phenotypically equivalent, $\Xi p_1^{(t)} = \Xi p_2^{(t)}$. Their projected offspring populations though are different: $\Xi p_1^{(t+1)} = \Xi\mathcal{M}(\cdot|g_1) \neq \Xi p_2^{(t+1)} = \Xi\mathcal{M}(\cdot|g_2)$. Hence, in the second case $\mathcal{M}$ is not compatible with phenotype equivalence. $\qquad\square$

Consider the following examples:

Example 1.2.1 *(Trivial GP-maps).* Assume that we encode (phenotypic) strings of length n by (genotypic) strings of length $n+1$, such that the GP-map simply ignores the genotype's last symbol. Obviously, this encoding is non-injective. But what is more important is that neutrality is trivial in that case, i.e., $\mathcal{G}$ commutes with Ξ. Hence, the encoding really is redundant—the additional bit has no effect whatsoever on phenotype evolution.

Many investigations that aim to argue against neutral encodings actually only investigate such trivial neutrality; they forget that neutrality (in their terms "redundancy") can have an absolute crucial impact on phenotypic variability in the case of a non-trivial GP-map.

In contrast, non-trivial neutrality is implicit in many models of evolutionary computation. Evolution strategies that make use of strategy parameters are the most basic paradigm, which we have already mentioned. But non-trivial neutrality occurs also in many other models. An excellent example are grammar-like genotype-phenotype mappings. Here, the same final phenotype can be represented by different sets of "developmental" rules. Depending on this representation, exploration in the space of phenotypes is very different. The following basic example anticipates the scenario in the computational model of section 1.5 and nicely demonstrates the variability of variational topology on the phenotype space.

Example 1.2.2 *(Non-trivial GP-maps).* Let the string abab be represented in the first case by the grammar {start-symbol→x, x→ab}, and in the second case by the grammar {start-symbol→abab}. If mutations are only rhs symbol flips, then the phenotypic variational neighbors of abab are, in the first case, {*, *b*b, a*a*} and in the second case {*bab, a*ab, ab*b, aba*}, where * means a symbol flip. These are quite different topologies!

1.2.7   Behind this formalism

The abstract formalism developed so far relates to topics like evolving genetic representations, strategy parameters, and evolvability.

There are no "fitness landscapes" over phenotype space.   In the literature, the notion of fitness landscapes is very often used as a metaphor to discuss and think about the fitness functions. If one refers to the fitness function over the *genotype* space, the idea is that the topology on the genotype space, given by the mutational variability $\mathfrak{M}(\cdot|g)$, allows to think of smooth and non-smooth function *w.r.t. this topology*, define local minima, and illustrate the function as a virtual landscape over G. However, the notion of fitness landscapes is sometimes also used to describe a fitness function over the *phenotype* space in the case of a non-trivial genotype-phenotype mapping. This is misleading because there exists no fixed topology on the phenotype space and thus no unique definition of local minima or smoothness; the phenotypic search points cannot be arranged w.r.t. to some meaningful neighborhood such the fitness function can be illustrated as a global landscape. In particular it seems counter-instructive to discuss the very effects of non-trivial genotype-phenotype mappings in terms of fitness landscapes over phenotype space, as it is though typically done: The non-fixed topology on P is then explained in terms of "shortcuts" or "bypasses" in the landscape that appear or disappear depending on the current position in a neutral set (Conrad 1990; Schuster 1996).

On evolving genetic representations.   The GP-map is commonly been thought of "the choice of representation". When applying evolutionary algorithms for problem solving, this choice of representation is usually crucial for the algorithm's efficiency. Even if one has two isomorphic representations, their effects on evolution may be very different since the algorithm's designer will usually define the mutation operators according to a seemingly natural topology on that representation (like the hypercube topology on bit strings). A common but misleading conclusion is that adaptability of exploration strategies requires an adaptability of the GP-map (Altenberg 1995; Wagner & Altenberg 1996). However, as our formalism clarifies, adaptability of exploration can be achieved for a *fixed*, but *non-trivial* genotype-phenotype mapping. In this case, a variation of exploration strategy does not occur by varying the GP-map but by neutral variations in the genotype space. For example, the genotype space may be considered very large, comprising *all* thinkable genetic representations of phenotypes (strings *and* trees *and* grammars, etc.). In that case, different neutral traits literally correspond to different genetic representations—such that a neutral variation allows for a change of representation although the genotype-phenotype map *as a whole* remains fixed. Of course, this presumes that there exist *neutral* mutations between these representations—in the case of the artificial evolution of strings, trees, and grammars, this is straightforward to realize and we will do so in section 1.5. In the case of natural evolution, the neutral transition from RNA to DNA genomes in early evolution is a corresponding paradigm (or think of the reverse transcription of an RNA sequence into a DNA sequence that allows the HI virus to insert its code in the human genome). In contrast, a self-adaptation of the whole genotype-phenotype mapping hardly makes sense since it is generally inconsistent to speak of a mapping from genotype to phenotype being parameterized by the genotype. We will reconsider this issue in section 1.4.4.

Strategy parameters as a special case. In (Toussaint & Igel 2002) we already discussed some implications of an exploration distribution $\mathcal{M}(\cdot|g)$ that depends on neutral traits (i.e., genotypic representations $g \in [x]$ in a given neutral set): It allows the exploration strategy (e.g., the variational topology) to *self-adapt*. In traditional approaches, in particular evolution strategies, so-called *strategy parameters* play the part of neutral traits—strategy parameters are a direct parameterization of mutation operators which are themselves part of the genotype (Angeline 1995; Smith & Fogarty 1997; Bäck 1998b). In these approaches, the genotype space is a Cartesian product $G = \tilde{G} \times Z$ of the phenotype space and the space Z of neutral strategy parameters. In some sense, our formalism generalizes the concept of strategy parameters to the case where the genotype space can not be decomposed into a product of phenotype and strategy parameter spaces but consists of arbitrarily interweaved neutral sets.

Evolvability.   All of this is strongly related to the discussion of *evolvability* in the biology literature. Many discussions are based on some relation between neutrality and evolvability (Schuster 1996; Kimura 1983) but to my knowledge there hardly exists a generic theoretical formalism to investigate such issues. (Analytical approaches to describe canalization, i.e., the evolution of mutational robustness of certain characters, based on neutral variations have been proposed by Rice (1998) and Wagner, Booth, & Bagheri-Chaichian (1997).) Following Wagner & Altenberg (1996), evolvability denotes the capability to explore further and further *good* individuals during evolution, which seems possible only when the exploration strategy itself is adapted in favor of evolvability during evolution. In our formalism this raises the question of how neutral traits, and thereby exploration distributions, do actually evolve. We will propose an answer to this question in section 1.2.9, which requires the formalism introduced in the next section.

1.2.8   Embedding neutral sets in the variety of exploration distributions

The embedding defines a unique way of understanding neutral traits and will allow to derive an evolution equation for them.

Our goal is to describe the evolution of neutral traits and thereby the evolution of exploration distributions $\mathcal{M}(\cdot|g)$. In simple cases where the genotype space decomposes, $G = X \times Z$, there seems no conceptual difficulty to do this: The evolutionary process as described on the genotype space may be projected on the hyperplane Z. This is the case for conventional strategy parameters and enabled Beyer (2001) to even derive analytical solutions for this process. However, since in general neutral traits live in arbitrarily interweaved neutral sets it seems tedious to find a description of their evolution. We are missing some embedding space to formulate equations—this actually reflects that we are missing a uniform way of interpreting and modeling neutral traits.

We now propose such an embedding. To simplify the notation let us "not distinguish" between two genotypes $g_1, g_2 \in [x]$ which induce the same exploration distribution $\mathcal{M}(\cdot|g_1) = \mathcal{M}(\cdot|g_2) \in \Lambda^G$. Formally, this means that we define another equivalence relation. Not distinguishing equivalent g's means considering only the evolution equation on the respective quotient space. It is clear the $\mathcal{M}$

commutes with this equivalence and of course also selection does. Since these
circumstances are rather obvious we skip introducing formal symbols and, from
now on, just assume that all g's in $[x]$ induce different distributions $\mathcal{M}(\cdot|g)$.

Thus, there exists an bijection between $[x]$ and the set

$$\overline{[x]} = \{\mathcal{M}(\cdot|g) \mid g \in [x]\} \subset \Lambda^G \tag{1.9}$$

of exploration distributions. It is this bijection that we want to emphasize because it defines an embedding of neutral sets in the space of exploration distributions. Specifically, it defines an embedding of the non-decomposable genotype
space G in a *product* space of "Phenotype × Distribution":

Definition 1.2.12 (σ-embedding (Toussaint 2003b)). Given a genotype
space G, a genotype-phenotype mapping $\phi : G \to P$, and a mutation operator
$\mathcal{M} : G \to \Lambda^G$, we define the σ-embedding as:

$$G \to P \times \Lambda^G \ : \ g \mapsto (x, \sigma) = \big(\phi(g), \mathcal{M}(\cdot|g)\big) . \tag{1.10}$$

The embedding space is $\bar{G} = P \times \Lambda^G$. Note that this mapping is injective (but
of course not surjective) and thus there exists a one-to-one relation between the
genotype space G and the subset $\{(x, \sigma) \mid x \in P, \ \sigma \in \overline{[x]}\} \subset \bar{G}$. The injectiveness
allows to directly associate a genotype distribution $p \in \Lambda^G$ with a distribution
over $\bar{G}$ by

$$p(x, \sigma) = \left\{ \begin{array}{ll} 0 & \text{if } \sigma \notin \overline{[x]} \\ p(g) & \text{if } \sigma \in \overline{[x]}, \ x = \phi(g), \text{ and } \sigma = \mathcal{M}(\cdot|g) . \end{array} \right. \tag{1.11}$$

The product structure of the embedding space is the key to formulate the evolution equation of exploration distributions in the next section. The embedding
also offers a new formal way of modeling neutral traits as specifying an exploration distribution $\sigma \in \overline{[x]}$; the neutral set $[x]$ being nothing but an isomorphic
copy of the space $\overline{[x]} \subset \Lambda^G$ of exploration distributions.

1.2.9 σ-evolution: A theorem on the evolution of pheno-
typic variability

*Exploration distributions naturally evolve towards minimizing the KL-divergence
between exploration and the exponential fitness distribution, and minimizing the entropy of exploration.*

To derive the theorem we neglect the stochastic elements, i.e., adopt the infinite population approach, and rewrite the general evolution equation (1.6, page 29) as

$$p^{(t+1)}(g') = \sum_{g \in G} \frac{f^{(t)}(g')}{\bar{f}^{(t)}} \, \mathcal{M}(g'|g) \, p^{(t)}(g) \, . \qquad (1.12)$$

We embed the equation in $\bar{G}$ and, according to equations (1.10) and (1.11), identify the exploration distribution $\mathcal{M}(x',\sigma'|g)$ with $\sigma(x',\sigma')$,

$$p^{(t+1)}(x',\sigma') = \sum_{x \in P} \sum_{\sigma \in \Lambda^G} \frac{\tilde{f}^{(t)}(x')}{\bar{f}^{(t)}} \, \sigma(x',\sigma') \, p^{(t)}(x,\sigma) \, .$$

This allows to run the summation over all possible distributions $\sigma \in \Lambda^G$; note that $\sigma \notin \overline{[x]} \Rightarrow p^{(t)}(x,\sigma) = 0$. We now benefit from $\bar{G}$ being a product space: The summations commute and executing summation over x gives

$$p^{(t+1)}(x',\sigma') = \sum_{\sigma} \frac{\tilde{f}^{(t)}(x')}{\bar{f}^{(t)}} \, \sigma(x',\sigma') \, p^{(t)}(\sigma) \, .$$

Here, $p^{(t)}(\sigma)$ is the marginal distribution of $p^{(t)}(x,\sigma)$, well defined because $\bar{G} = P \times \Lambda^G$ is a product space. Summing over x' and decomposing the mutation probability $\sigma(x',\sigma') = \sigma(x'|\sigma') \, \sigma(\sigma')$ we finally get

$$p^{(t+1)}(\sigma') = \sum_{\sigma} \frac{\sum_{x'} \tilde{f}^{(t)}(x') \, \sigma(x'|\sigma')}{\bar{f}^{(t)}} \, \sigma(\sigma') \, p^{(t)}(\sigma) \, .$$

We summarize this in

Theorem 1.2.3 (σ-evolution (Toussaint 2003b)). *Given the evolutionary process (1.12) on the genotype space G, the evolution of exploration distributions is described by the projection of the process on Λ^G given by the σ-evolution equation*

$$p^{(t+1)}(\sigma') = \sum_{\sigma} \frac{\langle \tilde{f}^{(t)}, \sigma(\cdot|\sigma') \rangle}{\bar{f}^{(t)}} \, \sigma(\sigma') \, p^{(t)}(\sigma) \, , \qquad (1.13)$$

where $\langle f, g \rangle := \sum_{x' \in P} f(x') \, g(x')$ denotes the scalar product in the function space L^2 over P.

σ-evolution describes the transition of a parent population $p^{(t)}(\sigma)$ of exploration distributions to the offspring population $p^{(t+1)}(\sigma')$ of exploration distributions. Therein, the term $\sigma(\sigma')$ corresponds to the mutation operator on Λ^G (recall that $\sigma(x', \sigma')$ corresponds to a mutation distribution $\mathcal{M}(x', \sigma'|g)$), and the equation matches the standard evolution equation in the form "$p^{(t+1)} = \mathcal{F} \mathcal{M} p^{(t)}$". In the following we discuss three aspects of the most interesting part of this equation, the fitness term $\left\langle \tilde{f}^{(t)} , \sigma(\cdot|\sigma') \right\rangle$.

The scalar product as quality measure. A term $\left\langle \tilde{f}^{(t)} , \sigma \right\rangle$ is a measure for σ that we call σ-*quality*. In the first place, it is the scalar product of $\tilde{f}^{(t)}$ with σ in the space of functions over P. The scalar product is a measure of the similarity and thus, σ-*quality measures the similarity between the exploration distribution and fitness.* σ-quality is very similar to the concept of effective fitness (Nordin & Banzhaf 1995; Stephens & Vargas 2000).

A delay effect. However, equation (1.13) exhibits that the term $\left\langle \tilde{f}^{(t)} , \sigma(\cdot|\sigma') \right\rangle / \tilde{f}^{(t)}$ is actually the fitness term for the *offspring* σ'. Thus, the fitness one has to associate with an offspring is the σ-quality *of its parent* under the condition that σ has in fact generated the offspring σ' ($\sigma(\cdot|\sigma')$ is the parent's phenotypic exploration distribution). Roughly speaking, the offspring is selected according to the quality of its parent. This circumstance can be coined a *first order delay* of evaluation. The quality of an individual's exploration distribution is not rewarded immediately by higher selection probability of this individual itself. It is rewarded when its offspring are selected with higher probability. This delay effect is well-known in the context of evolution strategies.[3] It is straightforward to generalize it to arbitrary *degrees of equivalence* leading to n-th order delay, see below.

The information theoretic interpretation. Since σ is actually a probability distribution we can also give an information theoretic interpretation. Let us

---

[3]For evolution strategies it is a common approach to *first* mutate the strategy parameters z before mutating the objective variables x according to the new strategy parameters (Schwefel 1995; Bäck 1998a). In our formalism this means that in equation (1.13) the evaluated distribution $\sigma(\cdot|\sigma') = \Xi\mathcal{M}(\cdot|x, z')$ is "similar" (with same strategy parameters z') to the offspring's distribution $\Xi\sigma'(\cdot) = \Xi\mathcal{M}(\cdot|x', z')$. However, the evaluated and the offspring's exploration distributions still differ significantly because they depend on the objective variables x and x', respectively. The real σ-quality of strategy parameters z' becomes evident only in the next generation in combination with the offspring's objective parameters x'.

introduce the *exponential fitness distribution* as

$$F^{(t)} = \frac{\exp \tilde{f}^{(t)}}{C^{(t)}} \;, \quad C^{(t)} = \sum_{x'} \exp \tilde{f}^{(t)}(x') \;.$$

One would also call $F^{(t)}$ the soft-max or Boltzmann distribution of $\tilde{f}^{(t)}$. The σ-quality can now be rewritten as

$$\left\langle \tilde{f}^{(t)}, \sigma \right\rangle + \ln C^{(t)} = \sum_{x'} \sigma(x') \ln F^{(t)}(x') = -D\big(\sigma : F^{(t)}\big) - H(\sigma) \;.$$

Hence, σ-quality is proportional to the negative of the divergence (see definition (1.3, page 26)) between the exploration distribution σ and the exponential fitness distribution $F^{(t)}$ minus the entropy of exploration. We summarize this in

Corollary 1.2.4 (σ-evolution (Toussaint 2003b)). *The evolution of explo-ration distributions (σ-evolution) naturally has a selection pressure towards*

- *minimizing the KL-divergence between exploration and the exponential fit-ness distribution, and*

- *minimizing the entropy of exploration.*

This also means that, assuming fixed entropy of σ, the exponential fitness distri-bution is a fixed point of σ-evolution that corresponds to the quasispecies (Eigen, McCaskill, & Schuster 1989). This sheds new light on what Eigen & Schuster (1977) found:

> "The single (molecular) species, however, is not the true target of selection. Eq. (10) tells us that it is rather the quasi-species, i.e., an organized combination which emerges via selection. As such it is selected against all other distributions." (Eigen & Schuster 1977)

.

We aimed at this interpretation of σ-evolution because it establishes a bridge to numerous approaches and discussions already present in the literature. First of all we understand the relation between self-adaptive σ-evolution and the de-terministic adaptation schemes of the exploration distribution in Estimation-of-Distribution Algorithms which we captured in terms of Generic Heuristic

Search (definition 1.2.2, page 27). Actually, σ-evolution realizes a similar kind of adaptation—minimization of the Kullback-Leibler divergence between exploration and fitness—but in a self-adaptive way.

Further, the question of how variational properties evolve has also been raised in numerous variations (pleiotropy, canalization, epistasis, etc.) in the biology literature. These discussions aim at understanding how evolution can handle to introduce correlations between phenotypic traits or mutational robustness or functional modularity in phenotypic exploration. Our answer is that variational properties evolve as to approximate the selection distribution. If, for example, certain phenotypic traits are correlated in the selection distribution F, then the Kullback-Leibler divergence decreases if these correlations are also present in the exploration distribution σ.

1.2.10   Appendix — n-th order equivalence and n-th order delay

How could evolution arrange to increase the probability for children to generate with high probability children that generate with high probability ...etc... children with high fitness?

How could we consider also higher order delays? Eventually, how could evolution arrange to *increase the probability for children to generate with high probability children that generate with high probability ...etc... children with high fitness.* We propose the following. What we considered up to now was the case when two genotypes g_1, g_2 are equivalent, $g_1 \equiv g_2$, because they share the same phenotype. As a result we found a first order delay of selection of neutral characters. Now consider the case when two genotypes are *2nd order equivalent* because their phenotypes are the same, $g_1 \equiv g_2$, *and* their *phenotypic* explorations are the same, $\Xi\sigma_1 = \Xi\sigma_2$, i.e. $\sigma_1 \hat{\equiv} \sigma_2$. Then, these two exploration distributions have, in fact, the same σ-quality and there is no immediate difference in selection between them. However, when writing equation (1.13) for *two* time steps we find

$$p^{(t+2)}(\sigma'') = \sum_{\sigma',\sigma} \frac{\left\langle \tilde{f}^{(t+1)}, \sigma'(\cdot|\sigma'')\right\rangle}{\bar{f}^{(t+1)}} \frac{\left\langle \tilde{f}^{(t)}, \sigma(\cdot|\sigma')\right\rangle}{\bar{f}^{(t)}} \sigma'(\sigma'')\,\sigma(\sigma')\,p^{(t)}(\sigma)\,.$$

Which means that there exists a longer term difference in selection between σ_1 and σ_2 iff the expected exploration distributions σ' *of their offspring* are not equivalent in the sense

$$\sum_{\sigma'} \sigma_1(\sigma')\,(\Xi\sigma') \neq \sum_{\sigma'} \sigma_2(\sigma')\,(\Xi\sigma') \; .$$

This difference in selection after two generations may be coined a second order delay effect. Generally, let us define two genotypes, (x_1, σ_1) and (x_2, σ_2), n-th order equivalent iff their phenotype and all their expected phenotypic exploration distributions after $0..n-2$ mutations are the same,

$$(x_1, \sigma_1) \equiv^n (x_2, \sigma_2) \quad \Longleftrightarrow$$
$$x_1 = x_2 \;, \quad n \geq 2 \Rightarrow \Xi\sigma_1 = \Xi\sigma_2 \;,$$
$$\forall_{1 \leq k \leq n-2} : \quad \sum_{\sigma^1,..,\sigma^k} \sigma_1(\sigma^1)\,\sigma^1(\sigma^2)..\sigma^{k-1}(\sigma^k)\,(\Xi\sigma^k)$$
$$= \sum_{\sigma^1,..,\sigma^k} \sigma_2(\sigma^1)\,\sigma^1(\sigma^2)..\sigma^{k-1}(\sigma^k)\,(\Xi\sigma^k) \;.$$

It follows that, if two genotypes are n-th order equivalent, then there exists no difference between the phenotypic dynamics of their evolution for the next $n-1$ generations. Thereafter though, with a delay of n generations, their evolution differs because the expected exploration distributions of the n-th order offspring of these two genotypes are not phenotypically equivalent.

1.2.11 Summary

Recalling the steps toward this result.

Why have we introduced Generic Heuristic Search? Our definition of Generic Heuristic Search makes explicit what we mean by "learning about a problem structure" on an abstract mathematical level: namely incorporating the information given by evaluations (in particular correlations between different solution features) in forthcoming search. GHS pinpoints that the crucial ingredient needed to realize this principle is the choice of the parameterization Φ of the exploration distribution.

Why have we applied Vose's formalism of compatibility? The literature is full of discussions on the purpose or relevance of neutrality or redundancy in genetic representations. There is a lack of a formal basis on which to ground these discussions. The approach to formalize neutrality by means of equivalence classes is not new—but the strict application of simple and fundamental arguments concerning the compatibility of $\mathcal{G}$ with phenotype equivalence allows to derive under which conditions neutrality influences phenotype evolution: Non-dependence of (projected) mutation on neutral traits is equivalent to compatibility of the evolution equation with phenotype equivalence, which we termed trivial neutrality.

Why have we introduced the σ-embedding of neutral traits? For a description of the evolution of neutral traits it is helpful to have an embedding space in which to formulate the evolution equation. In simple cases where the genotype space decomposes (strategy parameters) an embedding space of neutral sets is obvious. To generalize to arbitrary neutral sets and arbitrary genotype-phenotype mappings we introduced the σ-embedding, i.e., we embedded neutral sets in the space of probability distributions over the genotype space (*exploration distributions*).

Finally, what have we learned about σ-evolution? We derived an evolution equation for exploration distributions. The selection term can be interpreted as a fitness of exploration distributions that measures the (scalar product) similarity to the fitness function. In terms of information theory, σ-evolution minimizes the Kullback-Leibler divergence between the exploration distribution and the exponential (Boltzmann) fitness distribution, and minimizes the entropy of exploration. It describes the accumulation of information given by the fitness distribution into genetic representations.

1.3 Crossover, buildings blocks, and correlated exploration

In the previous section we ruled out the possibility of crossover although the discussion of crossover has a very strong tradition in the field of genetic algorithms. We now make up for this and discuss crossover w.r.t. correlated variability by comparing it to mutational variability and Estimation-Of-Distribution Algorithms (EDAs), i.e., Generic Heuristic Search.

1.3.1 Two notions of building blocks of evolution

The notion of building blocks as induced by crossover does not match the notion of functional phenotypic building blocks as it is induced by investigating correlated phenotypic variability.

In the realm of evolutionary computation the notion of building blocks has been developed in Holland's original works (Holland 1975; Holland 2000) to describe the effect of crossover. In that respect, building blocks are composed of genes with more or less linkage between them. This is one to one with the notion of schemata and eventually lead to the schema theories (also first developed in these papers) which describe the evolution of these building blocks.

Since crossover is a biologically inspired concept, Holland's notion of building blocks is also relevant in understanding natural evolution. In the biology literature though, there exists a second notion of building blocks which has quite a different connotation. As a paradigm we choose the following phenomenon.

Example 1.3.1 *(Drosophila's eyeless gene).* In their experiments, Halder, Callaerts, & Gehring (1995) forced the mutation of a single gene, called *eyeless gene*, in early ontogenesis of a Drosophila Melanogaster. This rather subtle genotypic variation results in a severe phenotypic variation: An additional functionally complete eye grows at some place it was not supposed to. Here, the notion of a building block refers to the eye as a functional module which can be grown phenotypically by triggering a single gene. In other words, a single mutation of a gene leads to a highly complex, in terms of physiological cell variables highly correlated phenotypic variation. Such properties of the genotype-phenotype mapping are considered as the basis of complex adaptation (Wagner & Altenberg 1996).

Besides the discussion of crossover in GAs and that of functional modularity in natural evolution, there is a third field of research that relates to the discussion of building blocks, namely Estimation-of-Distribution Algorithms that we discussed as examples for Generic Heuristic Search. The key of these algorithms is that they are capable to induce this *second* notion of building blocks. For instance, consider a dependency tree $y^{(t)}$ as parameterization of the search distribution $\Phi y^{(t)} \in \Lambda^P$ where the leaves encode the phenotypic variables. The offspring are generated by *sampling* this probabilistic model, i.e., by first sampling the root variable of the tree, then, according to the dependencies encoded on the links, sampling the root's successor nodes, etc. Now, if we assume that the dependencies are very strong, say, deterministic, it follows that a single variation at the root leads to a completely correlated variation of all leaves. Hence, we may define a set of leaves which, due to their dependencies, always vary in high correlation as a functional phenotypic module in the same sense as for the eyeless paradigm.

In the Evolutionary Computation community there are some discussions on the relation between EDAs and crossover GAs. Some argue that the essence of EDAs is that they can model the evolution of crossover building blocks (schemata) by explicitly encoding the linkage correlations that are implicit in the offspring distribution of crossover GAs (Shapiro 2003, Introduction). In that sense, EDAs are faster versions of crossover GAs; faster because EDAs actively analyze correlations in the selection distribution.

In this section we basically want to analyze the relation between crossover and correlated variability. Hence, this also means a discussion of the relation between Holland's notion of building blocks and this second notion we mentioned; and eventually also a discussion of the relation between EDAs and crossover GAs.

For instance, we will point out that crossover induces a correlation in the search distribution that can certainly be modeled by graphical models. However, the structure of these correlations is limited to the correlations that have already been present in the parent population. Crossover can only preserve certain (by the crossover mask determined) linkage correlations and never explore new correlated constellations; in total, it decreases the correlations in the search distribution. In contrast, EDAs can account for correlated variability by amplifying the correlations that are present in the parent population after selection. Here, amplifying means, for instance, to increase the mutual information in a distribution proportionally to the increase of entropy (which corresponds to mutative exploration)—similar to increasing the covariance of a Gaussian proportional to the standard deviation in order to preserve the structural correlations. In the case of the dependency tree, the scenario could be that the root variable induces the main entropy and changes its value such that directly depending variables change their values in high dependence of this change. The constellation of this set of variables might be new (has not been present in the parent population) and thus entropy is increased, but the dependencies and correlations between variables are preserved.

After we setup our formalism in the next section, section 1.3.3 and 1.3.4 will present some theorems on the structure of the search distribution after mutation and crossover. With structure we mean the correlational structure that we measure by means of mutual information. Many of our arguments will be based on the increase and decrease of mutual information in relation to increase or decrease of entropy in the search distribution. Section 1.3.5 finally defines the notion of correlated exploration and thereby pinpoints the difference between linkage correlations in crossover GAs and correlated variability in EDAs.

1.3.2 The crossover GA: Explicit definitions of mutation and crossover

In order to define crossover and derive results we assume that a genotype is composed of a finite number of genes and that crossover and mutation obey some constraints.

To investigate crossover we will make some additional assumptions about the evolutionary process that we generally introduced in equation (1.6, page 29). First, we will make crossover more explicit by notating it with its own operator $\mathcal{C} : \Lambda^G \to \Lambda^G$ that enters the evolution equation of a crossover GA via

$$p^{(t+1)} = \mathcal{S}^\mu \, \mathcal{F}^{(t)} \, \mathcal{S}^\lambda \, \mathcal{M} \, \mathcal{C} \, p^{(t)} \ ,$$

As we implicitly handled it already in the previous section, $\mathcal{M}$ accounts only for mutation instead of both, mutation and crossover.

Defining crossover requires that the genotype space G is composed of a fixed number of gene-spaces, $G = G^1 \times \cdots \times G^N$. The space G^i of alleles of the i-th gene may be arbitrary. Given this structure of the genotype space we define:

Definition 1.3.1 (Crossover). We define crossover as an operator $\Lambda^G \to \Lambda^G$ parameterized by a crossover mask distribution $c \in \Lambda^{\{0,1\}^N}$ over the space $\{0,1\}^N$ of bit-masks, where N is the number of loci (or genes) of a genome in G:

$$\mathcal{C} : \Lambda^G \to \Lambda^G , \quad (\mathcal{C}p)(x) = \sum_{x_0, x_1 \in G} \mathcal{C}(x|x_0, x_1) \, p(x_0) \, p(x_1) ,$$

$$\mathcal{C}(x|x_0, x_1) = \sum_{m \in \{0,1\}^N} c(m) \, [x = x_0 \otimes_m x_1] ,$$

where the i-th allele of the m-crossover-product $x_0 \otimes_m x_1$ is the i-th allele of the parent x_{m_i}, i.e., $(x_0 \otimes_m x_1)^i = (x_{m_i})^i$. A bracket expression $[A = B]$ equals 1 for $A = B$ and 0 for $A \neq B$. We only consider *symmetric* crossover, where $c(m) = c(\bar{m})$ and $\bar{m}$ is the conjugate of the bit-string m.

We also make additional assumptions on the mutation operator as given in the following definition:

Definition 1.3.2 (Simple Mutation). As before, general mutation is an operator $\mathcal{M} : \Lambda^G \to \Lambda^G$ defined by the conditional probability $\mathcal{M}(y|x)$ of mutating from $x \in G$ to $y \in G$, with $\mathcal{M}p = \sum_x \mathcal{M}(\cdot|x) \, p(x)$. A *typical* mutation operator fulfills the constraints of symmetry and component-wise independence:

a) $\mathcal{M}(y|x) = \mathcal{M}(x|y)$

b) $G = G^1 \times \cdots \times G^N \;\Rightarrow\; \mathcal{M}(x|y) = \prod_{i=1}^{N} \mathcal{M}^i(x^i|y^i)$

In the following we will refer to the *simple* mutation operator for which all component-wise mutation operators $\mathcal{M}^i$ are such that the probability of mutating from x to y is constant for $x \neq y$:

$$\forall i : \mathcal{M}^i = \mathcal{M}^* , \quad \forall x \neq y \in G^* : \mathcal{M}^*(x|y) = \frac{\alpha}{n} ,$$

$$\forall x \in G^* : \mathcal{M}^*(x|x) = 1 - \frac{\alpha \, (n-1)}{n} ,$$

where $n = |G^*|$ and $0 \leq \alpha \leq 1$ denotes the mutation rate parameter.

It is important to realize that, in our formalism, crossover and mutation are deterministic operators over the space of distributions. The stochasticity is

solely captured by the offspring sampling operator S^λ. Hence, when we will
derive statements about $\mathcal{M}$ and $\mathcal{C}$ in the following, they will not account for the
stochasticity of offspring sampling.

Finally, we use some further standard notations:

- We denote the marginals of a distribution $p \in \Lambda^G$, $G = G^1 \times \cdots \times G^N$, by
  $p^i \in \Lambda^{G^i}$.

- Accordingly, $H^i(p)$ denotes the entropy of the i-th marginal $H(p^i)$.

- And the mutual information is given by $I(p) = \sum_i H^i(p) - H(p)$ and the
  pair-wise mutual information by $I^{ij} = \sum_{a,b} p^{ij}(a,b) \ln \frac{p^{ij}(a,b)}{p^i(a)\,p^j(b)}$.

- Given some operator $\mathcal{U} : \Lambda^G \to \Lambda^G$ we will use the notation $\Delta_\mathcal{U} B = B(\mathcal{U}p) - B(p)$ to denote the difference of a quantity $B : \Lambda^G \to \mathbb{R}$ under
  transition, e.g., the quantity may be the entropy $H(p)$.

1.3.3 The structure of the mutation distribution

Mutation increases entropy and decreases mutual information.

This section derives a theorem that simply states that mutation increases entropy
and decreases mutual information. It is surprising how non-trivial it is to prove
this intuitively trivial statement.

**Lemma 1.3.1 (Entropy of component-wise mutation
(Toussaint 2003c)).** *Consider the component-wise simple mutation operator*
$\mathcal{M}^*$ *as given in definition 1.3.2. It follows that*

a)
$$\mathcal{M}^* p(x) = (1-\alpha)\,p(x) + \alpha\,\frac{1}{n}\,,$$

*which is a linear mixture between p and the uniform distribution ("$\frac{1}{n}$")
with mixture parameter α.*

b) *For every non-uniform population distribution p, the entropy of $\mathcal{M}^*p$ is
greater than the entropy of p,*

$$H(\mathcal{M}^*p) > H(p)\,.$$

Proof. a)

$$\mathcal{M}^* p(x) = \sum_y \mathcal{M}^*(x|y)\, p(y)$$

$$= \Big[\sum_y \frac{\alpha}{n}\, p(y)\Big] - \frac{\alpha}{n}\, p(x) + \Big(1 - \frac{\alpha\,(n-1)}{n}\Big)\, p(x)$$

$$= \frac{\alpha}{n} + (1-\alpha)\, p(x)\,.$$

b) We generally show that the entropy increases if you mix a distribution with the uniform distribution. We prove this by considering the first two derivatives of the entropy functional with respect to the mixture parameter α. Let

$$q(x) = (1-\alpha)\, p(x) + \frac{\alpha}{n}\,,$$

and recall $H(q) = -\sum_x q(x) \ln q(x)$ and $(X \ln X)' = X'((\ln X) + 1)$. It follows

$$\frac{\partial}{\partial \alpha} H(q) = -\sum_x \Big[-p(x) + \frac{1}{n}\Big](\ln q(x) + 1) = \sum_x \Big[p(x) - \frac{1}{n}\Big] \ln q(x)\,,$$

$$\frac{\partial}{\partial \alpha} H(q)\big|_{\alpha=1} = \sum_x \Big[p(x) - \frac{1}{n}\Big] \ln \frac{1}{n} = 0\,,$$

$$\frac{\partial^2}{\partial \alpha^2} H(q) = -\sum_x \frac{(p(x) - \frac{1}{n})^2}{q(x)} < 0 \quad \text{if } p \text{ is non-uniform.}$$

What we found is that (i) the entropy is maximal for the extreme case $\alpha = 1$ since its derivative w.r.t. α at this point vanishes (of course, this corresponds to the case where q becomes the uniform distribution) and (ii) the second derivative is always negative if p is non-uniform. Hence, the plot of H versus α is comparable to an upside-down parabola with maximum at $\alpha = 1$. It follows that for all $\alpha < 1$ (to the left of the maximum) the derivative $\frac{\partial}{\partial \alpha} H(q)$ is positive. Entropy continuously increases with α. And hence, for every $0 < \alpha \leq 1$ and every non-uniform population p, $H(\mathcal{M}^* p) > H(p)$. $\qquad\square$

Theorem 1.3.2 (Entropy and mutual information of mutation (Toussaint 2003c)). *Consider the simple mutation operator $\mathcal{M}(x|y) = \prod_i \mathcal{M}^*(x^i|y^i)$ as given in definition 1.3.2. If $p \in \Lambda^G$ is non-uniform it follows that entropy increases, $H(\mathcal{M}p) > H(p)$, and mutual information decreases, $I(\mathcal{M}p) < I(p)$.*

Proof. We first prove that the cross entropy decreases. Assuming only two genes, the compound mutation distributions reads

$$\mathcal{M}p(x,y) = (1-\alpha)^2\, p(x,y) + (1-\alpha)\,\alpha\, p(x)\,\frac{1}{n} + (1-\alpha)\,\alpha\,\frac{1}{n}\,p(y) + \alpha^2\,\frac{1}{n}\,\frac{1}{n}$$

$$= (1-\alpha)\left[(1-\alpha)\,p(x,y) + \alpha\,\frac{1}{n}\,p(x)\right] + \alpha\,\frac{1}{n}\left[(1-\alpha)\,p(y) + \alpha\,\frac{1}{n}\right]$$

$$= (1-\alpha)\,q(x,y) + \alpha\,\frac{1}{n}\,q(y)\ ,$$

$$\text{where}\quad q(x,y) = (1-\alpha)\,p(x,y) + \alpha\,p(x)\,\frac{1}{n}\ ,$$

$$q(x) = p(x)\ ,\quad q(y) = (1-\alpha)\,p(y) + \frac{\alpha}{n}$$

We call q a one-component α-mixture since only in one component the uniform distribution was mixed to p. This shows that the compound distribution $\mathcal{M}p$ for two genes is a one-component α-mixture of a distribution q, which is itself a one-component α-mixture. For compound distributions with more than two genes this will be recursively the case and generally the mutation operator can be expresses as concatenation of one-component α-mixtures. Hence, it suffices when we prove that the mutual information decreases for one such step of one-component α-mixing.

We use the same technique of calculating derivatives with respect to the mixture parameter to proof decreasing cross entropy. To simplify the notation we use the abbreviations:

$$A = q(x,y)\ ,\ A\big|_{\alpha=1} = \frac{\alpha\,p(x)}{n}\ ,\ A' = \frac{\partial}{\partial\alpha}A = -p(x,y) + \frac{p(x)}{n}\ ,\ A'' = 0\ ,$$

$$B = q(x)\,q(y) = p(x)\left[(1-\alpha)\,p(y) + \frac{\alpha}{n}\right]\ ,\ B\big|_{\alpha=1} = A\big|_{\alpha=1}\ ,$$

$$B' = p(x)\left(-p(y) + \frac{1}{n}\right)\ ,\ B'' = 0\ .$$

With these abbreviations (keeping the dependencies on x, y, and α in mind) we can write:

$$I(q) = \sum_{x,y} A \ln \frac{A}{B}$$

$$\frac{\partial}{\partial\alpha}I(q) = \sum_{x,y}\left[A' \ln \frac{A}{B} + A' - \frac{A\,B'}{B}\right]$$

$$\frac{\partial}{\partial \alpha} I(q)\Big|_{\alpha=1} = \sum_{x,y} \left[A'\big|_{\alpha=1} \ln \frac{A\big|_{\alpha=1}}{A\big|_{\alpha=1}} + \left[-p(x,y) + \frac{p(x)}{n} \right] \right.$$

$$\left. - \frac{A\big|_{\alpha=1}}{A\big|_{\alpha=1}} p(x)\left(-p(y) + \frac{1}{n}\right) \right] = 0$$

$$\frac{\partial^2}{\partial \alpha^2} I(q) = \sum_{x,y} \left[A' \frac{B}{A}\left[\frac{A'}{B} - \frac{A B'}{B^2}\right] + 0 - \frac{A' B'}{B} + \frac{A (B')^2}{B^2} \right]$$

$$= \sum_{x,y} \left[\frac{(A')^2}{A} - 2\frac{A' B'}{B} + \frac{A (B')^2}{B^2} \right]$$

$$= \sum_{x,y} \left[\frac{(B A' - A B')^2}{A B^2} \right] \geq 0$$

So, what we found is that (i) for $\alpha = 1$ the cross entropy is minimal since its derivative w.r.t. α at this point vanishes (of course, this corresponds to the case where $q(x,y) = p(x)\frac{1}{n}$) and (ii) for all other points the second derivative is positive. The plot of I versus α is comparable to an upwards parabola with minimum at $\alpha = 1$. It follows that for $\alpha < 1$ (to the left of the minimum) the derivative $\frac{\partial}{\partial \alpha} I(q)$ is negative and thus the cross entropy continuously decreases with increasing α.

Concerning increasing entropy, it is obvious that the marginals of the mutation distribution $\mathcal{M}p$ are simply

$$(\mathcal{M}p)^i = \mathcal{M}^* p^i .$$

For the component-wise mutation operators we proved that entropy increases (for non-zero α and non-uniform p) and thus $\Delta_{\mathcal{M}} H^i > 0$. Consequently,

$$\Delta_{\mathcal{M}} H = \sum_i \Delta_{\mathcal{M}} H^i - \Delta_{\mathcal{M}} I > 0 .$$

□

1.3.4   The structure of the crossover distribution

Crossover destroys mutual information in the parent population by transforming it into entropy in the crossed population.

What is the structure of the crossover search distribution $\mathcal{C}p$, given the population $p \in \Lambda^G$ and the crossover mask distribution $c \in \Lambda^{\{0,1\}^N}$? The first theorem can directly be derived from our definition of the crossover operator. It captures the most basic properties of the crossover operator with respect to the correlations it *destroys* in the search distribution:

Theorem 1.3.3 (Entropy and mutual information of crossover (Toussaint 2003c)). *Let $H(p)$, p^i, $H^i(p) = H(p^i)$, and $I(p) = \sum_i H^i(p) - H(p)$ denote the entropy, the i-th marginal distribution, the marginal entropies, and the mutual information of a distribution p. For any crossover operator $\mathcal{C}$ and any population p it holds*

a) $\forall i: (\mathcal{C}p)^i = p^i$, $\Delta_{\mathcal{C}} H^i = 0$, *i.e., the marginals and hence their entropies do not change,*

b) $\Delta_{\mathcal{C}} I = -\Delta_{\mathcal{C}} H \leq 0$, *i.e., the increase of entropy is equal to the decrease of mutual information.*

Proof. Let us first calculate the marginals after crossover. Let a be an allele of the i-th gene.

$$(\mathcal{C}p)^i(a) = \sum_{x_0, x_1} \sum_m c(m) \left[a = (x_{m_i})^i \right] p(x_0) \, p(x_1) ,$$

$$= \sum_{x_0, x_1} \left[\sum_{m:m_i=0} c(m) \left[a = (x_0)^i \right] + \sum_{m:m_i=1} c(m) \left[a = (x_1)^i \right] \right] p(x_0) \, p(x_1)$$

$$= p^i(a) \left[\sum_{m:m_i=0} c(m) \right] + p^i(a) \left[\sum_{m:m_i=1} c(m) \right] = p^i(a) .$$

Since the marginals are not changed by crossover, the marginal entropies do not change either. Statement *b)* follows from the definition of the mutual information:

$$\Delta_{\mathcal{C}} H + \Delta_{\mathcal{C}} I = H(\mathcal{C}p) - H(p) + I(\mathcal{C}p) - I(p)$$

$$= H(\mathcal{C}p) - H(p) + \sum_i H^i(\mathcal{C}p) - H(\mathcal{C}p) - \sum_i H^i(p) + H(p)$$

$$= \sum_i H^i(\mathcal{C}p) - \sum_i H^i(p) = 0 .$$

$\square$

The following theorem makes this more concrete when focusing on two specific genes of a genome of arbitrary length. We calculate the mutual information between these two genes in the search distribution $\mathcal{C}p$—which is a measure for the *linkage* between them. Let it be the i-th and j-th gene. We use a and b as alleles; $p^{ij}(a,b) = \sum_{x \in G} [x^i = a] \, [x^j = b] \, p(x)$ denotes the probability that the i-th gene has allele a and the j-th gene allele b. Analogously, let c^{ij} be the marginal of the crossover mask distribution with respect to the two genes, i.e., $c^{ij}(01) = \sum_{m \in \{0,1\}^N} [m^i = 0] \, [m^j = 1] \, c(m)$.

Theorem 1.3.4 (Two-gene entropy and mutual information of crossover (Toussaint 2003c)). *For any crossover operator $\mathcal{C}$ and any population p it holds:*

a) *The compound distribution of two genes after crossover is given by*

$$(\mathcal{C}p)^{ij}(a,b) = 2 \, c^{ij}(00) \, p^{ij}(a,b) + 2 \, c^{ij}(01) \, p^i(a) \, p^j(b) \,,$$

i.e., a linear combination of the original compound distribution $p^{ij}(a,b)$ and the decorrelated product distribution $p^i(a) \, p^j(b)$.

b) *The mutual information $I(\mathcal{C}p)^{ij}$ in the compound distribution of two specific genes is*

$$I(\mathcal{C}p)^{ij} = \sum_{a,b} (\mathcal{C}p)^{ij}(a,b) \ln \left(2c^{ij}(00) \frac{p^{ij}(a,b)}{p^i(a)p^j(b)} + 2c^{ij}(01) \right) ,$$

c) *and we have*

$$0 \;\leq\; 2c^{ij}(00) \left(I(p)^{ij} + \ln(2c^{ij}(00)) \right) \;\leq\; I(\mathcal{C}p)^{ij} \;\leq\; I(p)^{ij} \,.$$

The two left $\leq$ are exact for complete crossover, $c^{ij}(00) = 0$, $c^{ij}(01) = \frac{1}{2}$, the right $\leq$ is exact for no crossover, $c^{ij}(00) = \frac{1}{2}$, $c^{ij}(01) = 0$.

Proof. a)

$$\mathcal{C}p^{ij}(a,b) = \sum_{x_0,x_1} \sum_m c(m) \, [(x_{m_0})^0 = a] \, [(x_{m_1})^1 = b] \, p(x_0) \, p(x_1)$$

$$= \sum_{x_0,x_1} \Big(c^{ij}(00) \, [(x_0)^0 = a][(x_0)^1 = b] + c^{ij}(01) \, [(x_0)^0 = a][(x_1)^1 = b]$$

$$+ c^{ij}(10) \left[(x_1)^0 = a\right]\left[(x_0)^1 = b\right] + c^{ij}(11) \left[(x_1)^0 = a\right]\left[(x_1)^1 = b\right]\Big)$$

$$\cdot\, p(x_0)\, p(x_1)$$

$$= 2 \sum_{x_0} c^{ij}(00) \left[(x_0)^0 = a\right]\left[(x_0)^1 = b\right] p(x_0)$$

$$+ 2 \sum_{x_0,x_1} c^{ij}(01) \left[(x_0)^0 = a\right]\left[(x_1)^1 = b\right] p(x_0)\, p(x_1)$$

$$= 2\, c^{ij}(00)\, p^{ij}(a,b) + 2\, c^{ij}(01)\, p^i(a)\, p^j(b)\ .$$

b&c)

$$I(\mathcal{C}p)^{ij} = H(\mathcal{C}p^i) + H(\mathcal{C}p^j) - H(\mathcal{C}p) = H(p^i) + H(p^j) - H(\mathcal{C}p)$$

$$\leq H(p^i) + H(p^j) - H(p) = I(p)^{ij}$$

$$H(\mathcal{C}p) = -\sum_{a,b}(\mathcal{C}p)^{ij}(a,b) \ln\left(2\, c^{ij}(00)\, p^{ij}(a,b) + 2\, c^{ij}(01)\, p^i(a)\, p^j(b)\right)$$

$$= -\sum_{a,b}(\mathcal{C}p)^{ij}(a,b)\left[\ln\left(2c^{ij}(00)\frac{p^{ij}(a,b)}{p^i(a)p^j(b)} + 2c^{ij}(01)\right)\right.$$

$$\left. - \ln p^i(a) - \ln p^j(b)\right]$$

$$= -\sum_{a,b}(\mathcal{C}p)^{ij}(a,b)\left[\ln\left(2c^{ij}(00)\frac{p^{ij}(a,b)}{p^i(a)p^j(b)} + 2c^{ij}(01)\right)\right]$$

$$+ H(p^i) + H(p^j)$$

$$I(\mathcal{C}p)^{ij} = \sum_{a,b}\left(2\, c^{ij}(00)\, p^{ij}(a,b) + 2\, c^{ij}(01)\, p^i(a)\, p^j(b)\right)$$

$$\cdot \ln\left(2c^{ij}(00)\frac{p^{ij}(a,b)}{p^i(a)p^j(b)} + 2c^{ij}(01)\right)$$

$$\geq \sum_{a,b}\left(2\, c^{ij}(00)\, p^{ij}(a,b)\right) \ln\left(2c^{ij}(00)\frac{p^{ij}(a,b)}{p^i(a)p^j(b)}\right)$$

$$= 2c^{ij}(00)\left(I(p)^{ij} + \ln(2c^{ij}(00))\right)$$

$\square$

Let us summarize what we actually found in the above theorems:

- The marginal distributions do not change at all. There is no exploration w.r.t. the alleles of single genes.

- The more entropy crossover introduces in a population, the more the mutual dependencies between genes are destroyed. Actually, crossover destroys mutual information in the parent population by *transforming* it into entropy in the crossed population. In particular, if there is no mutual information in the parent population, crossover will not generate any more entropy. That's linkage equilibrium.

- The last theorem shows how the crossover mask distribution c determines *which* correlations are destroyed and transformed into entropy.

The purpose of these theorems is to propose a probably non-standard point of view on what crossover actually does: Actually, a *non*-crossover GA comprises the strongest and most natural building blocks; individuals as such are the building blocks that carry the mutual information between their genes. Crossover is a means to break these maximal building blocks apart into smaller pieces by converting mutual dependencies into entropy. As a result it induces smaller, more fine-grained building blocks with, in total, less mutual information in the crossed population. Hence, the correlational structure in the crossed population is not more complex—it is simpler since it carries less information. In the limit of linkage equilibrium (or uniform c), all correlations have been destroyed and the crossed population becomes a product distribution.

1.3.5   Correlated exploration

A precise definition of correlated exploration allows to pinpoint the difference between crossover exploration and correlated exploration in the case of EDAs.

Both, crossover and EDAs have a non-trivial influence on the correlational structure in the search distribution. The crucial difference is that Estimation-of-Distribution Algorithms try to "carry over" the correlations in the population of selected to the search distribution (cf. the *estimation* in equation (1.4, page 27)) whereas crossover destroys correlations. Carrying over correlations is non-trivial

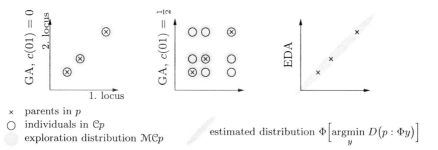

×   parents in p
○   individuals in $\mathcal{C}p$
     exploration distribution $\mathcal{MC}p$ estimated distribution $\Phi\left[\operatorname*{argmin}_{y} D(p : \Phi y)\right]$

Figure 1.1: Illustration of the type of correlations in GAs with and without crossover in comparison to correlated exploration in EDAs (cf. definition 1.2.2, page 27). The gray shades indicate the exploration distributions, say, regions of probability greater than some constant. The degree to which the gray shading is aligned with the bisecting line indicates correlatedness. The crossover GA in the middle destroys correlations whereas EDAs may induce high correlations. We assumed that the loci can have several different values—in the case of bit-strings these loci could refer to several bits and $c(01)$ denoted the probability for 1-point crossover between these groups of bits.

if the search distribution is to be explorative, i.e., of more entropy: Typical mutation operators add entropy to the distribution by adding independent noise to each marginal, but this reduces the mutual information between genes (see Lemma 1.3.2).

Consider illustration 1.1. In a finite population of 3 individuals, marked by crosses, the values at the two loci are correlated, here illustrated by plotting them on the bisecting line. The crossed population $\mathcal{C}p$ comprises at most 9 different individuals; in the special cases $c^{ij}(01) = 0$ and $c^{ij}(01) = \frac{1}{2}$ the population is even finite and comprises 3 respectively 9 equally weighted individuals marked by circles. Mutation adds independent noise, illustrated by the gray shading, to the alleles of each individual. The two illustrations for the GA demonstrate that crossover destroys correlations between the alleles in the initial population instead of carrying it over to the search distribution: The gray shading is not focused on the bisecting line. Instead, an EDA would first estimate the distribution of the individuals in p. Depending on what probabilistic model is used, this model can capture the correlations between the alleles; in the illustration the model could be a Gaussian parameterized by the mean and covariance matrix (just as for the CMA evolution strategy (Hansen & Ostermeier 2001)) and the estimation of the correlations in p leads to the highly structured search distribution in which the entropy of each marginal is increased without destroying the correlations between them. We capture this difference in the following definition:

Definition 1.3.3 (Correlated exploration (Toussaint 2003c)). Let $\mathcal{U}$: $\Lambda^G \to \Lambda^G$ be an operator. The following conditions need to hold for *almost all* $p \in G$ which means for all the space G except for a subspace of measure zero. We define

- $\mathcal{U}$ is explorative $\iff \Delta_\mathcal{U} H > 0$ for almost all $p \in G$,

- $\mathcal{U}$ is marginally explorative $\iff \mathcal{U}$ is explorative and $\exists i : \Delta_\mathcal{U} H^i > 0$ for almost all $p \in G$,

- $\mathcal{U}$ is correlated explorative $\iff \mathcal{U}$ is explorative and $\Delta_\mathcal{U} I > 0$, or equivalently $0 < \Delta_\mathcal{U} H < \sum_i \Delta_\mathcal{U} H^i$, for almost all $p \in G$.

Corollary 1.3.5 (Correlated exploration, mutation, and crossover (Toussaint 2003c)). *From this definition it follows that*

a) *If and only if there exist two loci i and j such that the marginal crossover mask distribution $c^{ij}(01)$ for these two loci is non-vanishing, $c^{ij}(01) = c^{ij}(10) > 0$, then crossover $\mathcal{C}$ is explorative. For every mask distribution $c \in \Lambda^{\{0,1\}^N}$, crossover $\mathcal{C}$ is neither marginally nor correlated explorative.*

b) *Mutation $\mathcal{M}$ is marginally but not correlated explorative.*

c) *Mutation and crossover $\mathcal{M} \circ \mathcal{C}$ are marginally but not correlated explorative.*

d) *In the case of a non-trivial genotype-phenotype mapping mutation as well as crossover can be* phenotypically *correlated explorative.*

Proof. a) That $\mathcal{C}$ is neither marginally nor correlated explorative follows directly from Theorem 1.3.3a, which says that for every $c \in \Lambda^{\{0,1\}^N}$ and any population $p \in \Lambda^G$ the marginals of the population do not change under crossover, $\Delta_\mathcal{C} H^i = 0$. But under which conditions is $\mathcal{C}$ explorative?

If, for two loci i and j, $c^{ij}(01)$ is non-vanishing, it follows that $\mathcal{C}$ reduces the mutual information between these two loci (Theorem 1.3.4c). The subspace of populations p that do not have any mutual information I^{ij} between these

two loci is of measure zero. Hence, for almost all p, $\Delta_{\mathrm{C}} I^{ij} < 0$ and, following
Theorem 1.3.3b this automatically leads to an increase of entropy $\Delta_{\mathrm{C}} H^{ij} > 0$ in
the compound distribution of the two loci and, since $\Delta_{\mathrm{C}} H \geq \Delta_{\mathrm{C}} H^{ij}$, also of the
total entropy.

The other way around, if, for every two loci i and j, $c^{ij}(01)$ vanishes it follows
that there is no crossover, i.e., only the all-0s and all-1s crossover masks have
non-vanishing probability. Hence, $\mathcal{C} = \mathrm{id}$ and is not explorative.

b) In lemma 1.3.2 we prove that for every non-uniform population p $\Delta_{\mathrm{M}} H > 0$,
$\Delta_{\mathrm{M}} H^i > 0$, and $\Delta_{\mathrm{M}} I < 0$.

c) Since both, mutation and crossover are not correlated explorative, it follows
that their composition is also not correlated explorative:

$$\Delta_{\mathrm{C}} I \leq 0 \, , \ \Delta_{\mathrm{M}} I \leq 0 \quad \Rightarrow \quad \Delta_{\mathrm{MC}} I \leq 0 \, .$$

d) What is different in the case of a non-trivial genotype-phenotype mapping?
The assumptions we made about the mutation operator (component-wise independence) refer to the genotype space, not to the phenotype space: On genotype space mutation kernels are product distributions and mutative exploration
is marginally explorative but not correlated; projected on phenotype space, the
mutation kernels are in general not anymore product distributions and hence
phenotypic mutative exploration can be correlated. The same arguments hold
for crossover.   □

1.3.6   Conclusions

*Crossover is good to decorrelate exploration; is does not, as EDAs, induce
complex exploration.*

The evolutionary process, as given in equation (1.6, page 29) is a succession
of increase and decrease of entropy in the population. The fitness operator
adds information to the process by decreasing the entropy (it typically maps a
uniform finite distribution on a non-uniform with same support). And crossover
and mutation add entropy in order to allow for further exploration. In this
section we analyzed the structure of this explorative entropy, namely the mutual
information (cross entropy) inherent in the offspring distribution induced by

mutation and crossover. We showed that both, mutation and crossover are explorative but thereby destroy mutual information, i.e., they destroy structural information given by selection. In contrast, EDAs try to preserve or even amplify this structural information. Such amplification occurs, e.g., if one adds entropy to the distribution while trying to preserve the relative dependencies between variables (preserving correlations amplifies covariances in a Gaussian). This is what we defined correlated exploration. Crossover does the *inverse* of correlated exploration. It destroys mutual information in the exploration distribution by transforming it into entropy.

Of course, the crossover exploration distribution can be modeled by graphical models. In that respect, one could certainly design search algorithms based on probabilistic models of the search distribution that model crossover GAs—PBIL (Baluja 1994) is a candidate. However, one should not call such an algorithm an Estimation-of-Distribution Algorithm because its objective is not to really estimate the distribution of selected and in particular the correlations within this distribution. (The PBIL is an exception since its objective is to only estimate the marginals which coincides with modeling crossover). In general, EDAs go beyond modeling crossover since they introduce a quality which is not a quality of crossover: correlated exploration.

Finally, there is a crucial difference between EDAs and (crossover) GAs with respect to the self-adaptation of the exploration distribution. EDAs always adapt their search distribution (including correlations) according to the distribution of previously selected solutions. In contrast, the crossover mask distribution, that determines where correlations are destroyed or not destroyed, is usually not self-adaptive.

1.4 Rethinking natural and artificial evolutionary adaptation

The idea of this section is to review some basic phenomena of natural evolution and a few interesting artificial models in the language we developed in the previous section. The discussion will emphasize the abstract principles that seem relevant from the perspective of our theory while neglecting molecular details. One goal is to shed new light on these natural phenomena; the other is to clarify how the abstract theorems we developed relate to nature and increase intuition about them.

1.4.1 From the DNA to protein folding: Genotype vs. phenotype variability in nature

Without referring to its phenotypical meaning, the DNA is of a rather simple structure and variational topology. The codon translation is the first step in the GP-map that introduces neutrality and non-trivial phenotypic variability. Considering protein functionality as a phenotypic level, the GP-map from a protein's primary to tertiary structure is complicated enough that neutral sets percolate almost all the primary structure space.

Genotypic variability on the DNA. A mathematician should appreciate the way nature organizes complex variability on the phenotype level: A comparatively simple variational topology on the genotype space induces, by virtue of a complex genotype-phenotype mapping, a complex variational topology on the phenotype space. One may compare this to a homomorphism that maps from a simple structured base space (like the topological $\mathbb{R}^n$ or the Euclidean $\mathbb{R}^n$) to another space where it induces a more complex structure (like a topological manifold or a Riemannian manifold). In this section we briefly describe what we call "comparatively simple variational topology on the genotype space" before we discuss some aspects of this "complex genotype-phenotype mapping".

From an abstract point of view, the DNA (deoxyribonucleic acid) itself is of a rather simple abstract structure, organized as one or several sequences (Chromosomes) of nucleic acids (base pairs)—essentially a sequence over a 4-ary al-

phabet $\{A,T,G,C\}$ (Adenine, Thymine, Guanine, Cytosine). The major part of
genotype variability is induced by base pair replacements, see table 1.1. Such
mutations correspond to a variational topology similar to a hypercube, where
in each dimension the four points A,T,G,C are fully connected. In particular,
there are no correlations, no complex structure present in such kind of variability (think of a Euclidean metric in the base space of a manifold). Insertion
and deletion of base pairs resemble a similarly kind of topology in the union
space $\bigcup_{n=1}^{\infty}\{A,T,G,C\}^{n}$ of all DNA sequences of any length. We call base pair
replacement, insertion, and deletion *1st-type mutations* and they constitute our
basic picture of variational topology on the DNA space. Of course, one needs to
point out that this topological interpretation of variability simplifies quantitative
aspects of mutation probabilities.[4]

There are also more complex types of mutations that we call *2nd-type mutations*.
Errors during replication or recombination of Chromosomes may lead to translo-
cations, deletions, or inversions of whole subsequences of DNA. For these types
of mutation, the topological interpretation hardly increases intuition. We will
not further discuss the meaning of 2nd-type mutations until, in section 1.5, a
new model will naturally lead to the necessity of similar "structural" mutations.
[5]

What we conclude is the following: If in nature the source of entropy, i.e., the
source of variability and exploration, are largely simple non-structured (non-
correlated) mutations of the genotype, then the only way nature could real-

---

[4]In fact, nature has invented several mechanisms to modulate variation probabilities at
specific loci, e.g.: Concerning 1st-type mutations, there exist so-called hot spots of spontaneous mutations. For instance, at position 104 of the lac I gene of the Escherichia coli
bacterium (E. coli) there is the sequence CCAGG. The second Cytosine element is, due to
its nucleotide neighbors, chemically modified to 5-Methylcytosine. It is a likely reaction
from 5-Methylcytosine to Thymine, which is not repaired by DNA repair mechanisms because Thymine is a regular nucleotide. Thus, the mutation $C \rightarrow T$ occurs frequently at this
particular position.

[5]Concerning 2nd-type mutations there exist several mechanisms that influence the probability of where such mutations occur: Transposons are mobile parts of the genome, sometimes
also called *jumping genes*, that are embraced by inverted repeats, i.e., identical sequences
reading in opposite. An enzyme that is encoded on the transposon itself (the transposase)
cuts the DNA strand at the inverted repeats and reinserts the transposon at another place
of the DNA (depending on where the transposase can bind to the DNA). Sometimes the
transposon is not cut out but replicated before it is reinserted (retrotransposons). A good
example is the speckled corn: A mutation in some cells during early development is caused by
a typical transposon translation and leads to a change of color; the corn becomes black. This
mutation is passed on to all the descendant cells and the corn becomes speckled. Actually
also the HI virus is similar to a retrotransposon: Its RNA encodes the necessary enzymes to
translate a RNA strand into a DNA sequence and insert it in the human DNA just like a
transposon is reinserted.

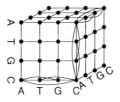

nucleic acid replacement	70.8%
deletion	17.2 %
insertion	7.7 %
frame shift mutation	4.3 %

genome type	mutability α	selective advantage f	max stable length $l_{\max}$
RNA without enzyme reproduction	$5 \cdot 10^{-2}$	2/20/200	14/60/106
RNA with enzyme reproduction	$5 \cdot 10^{-4}$	2/20/200	1386/5991/10597
DNA with enzyme reproduction	$1 \cdot 10^{-6}$	2/20/200	$0.7 \cdot 10^6/3 \cdot 10^6/5.3 \cdot 10^6$
Prokaryotes, E. coli $(l{=}4 \cdot 10^6)$	$1 \cdot 10^{-7}$	2/20/200	$0.7 \cdot 10^7/3 \cdot 10^7/5.3 \cdot 10^7$
Eukaryotes, human $(l{=}3 \cdot 10^9)$	$1 \cdot 10^{-9}$	2/20/200	$0.7 \cdot 10^9/3 \cdot 10^9/5.3 \cdot 10^9$

Table 1.1: The cube in top left illustrates the topology of single base pair replacement mutations for three base pairs (e.g. a codon). The table to the top right gives an idea of the relative frequencies of different kinds of mutations; it presents the statistics of spontaneous mutations in the lac I gene of E. coli (from Knippers 1997, page 242). The bottom table relates mutation rates to different genome types that occurred during natural evolution. The selective advantage is related to a quasi-species' reproduction rate relative to that of competitors; $l_{\max}$ gives the maximal genome length that allows for a stable quasi-species with given mutability and selective advantage, $l_{\max} = \frac{\ln f}{\alpha}$; see (Eigen & Schuster 1977) and (Schuster & Sigmund 1982) for details.

ize complex adaptation (a structured exploration distribution) is via a complex genotype-phenotype mapping.

Transcription and translation. If the DNA itself is a mere point in the space $\{A,T,G,C\}^n$ of simply structured, hypercube-like variational topology, we now need to discuss how complex DNA becomes *if* referring to its phenotypical meaning. In particular, the simple variational topology on DNA space translates to an incredibly complex structured variability on phenotype space.

The DNA might be compared to an operator Π in a dynamic system $\dot{x} = \Pi(x)$; it specifies how molecular concentrations change in a closed, cellular environment. In this view, a gene (i.e., a DNA-subsequence ⟨promoter, start codon, coding

	· U ·		· C ·		· A ·		· G ·	
U · ·	UUU	Phe	UCU	Ser	UAU	Tyr	UGU	Cys
	C		C		C		C	
	UUA	Leu	A		UAA	ochre stop	UGA	opal stop
	G		G		UAG	amber stop	UGG	Trp
C · ·	CUU	Leu	CCU	Pro	CAU	His	CGU	Arg
	C		C		C		C	
	A		A		CAA	Gln	A	
	G		G		G		G	
A · ·	AUU	Ile	ACU	Thr	AAU	Asn	AGU	Ser
	C		C		C		C	
	A		A		AAA	Lys	AGA	Arg
	AUG	Met	G		G		G	
G · ·	GUU	Val	GCU	Ala	GAU	Asp	GGU	Gly
	C		C		C		C	
	A		A		GAA	Glu	A	
	G		G		G		G	

Table 1.2: The codon translation table (from Knippers, 1997, page 76):
Most codons that code for the same amino acids have the first two nucleotides fixed
and the third arbitrary (with multiplicity 4) or varying over U,C or A,G (multiplicity
2). The only exceptions are the codons for Leucine, Arginine, and Serine for which the
first two nucleotides are not fixed (each with multiplicity 6); Ile, for which the third
varies over U,C,A (multiplicity 3); and Met and Trp, for which the third is fixed to G
(multiplicity 1). Note that these neutral sets are typically a neighborhood w.r.t. base
pair replacement.
An exception is the 21st amino acid Selenocysteine, present in some enzymes of bac-
teria and mammals, which is encoded by the opal stop codon UGA *iff* the sequence
neighborhood fulfills certain conditions.

sequence, stop codon) that codes for exactly one molecule) is one additive component of this operator. If Π *were* linear, the promoter would specify the column
and the coding sequence the row of the matrix element that corresponds to the
gene.

But the way how the coding sequence, written in letters {A,T,G,C}, represents
the molecule is highly non-trivial. After the gene is expressed, transcribed to the
messenger RNA, and this mRNA is further processed and spliced, triplets of this
alphabet (so-called codons) are mapped to amino acids.[6] Besides the fascinating

---

[6]Depending on existing molecule concentrations (transcription factors) and corresponding
gene promoters, enzymes (RNA polymerase) bind to the genes and start *transcription*: The

biological implementation of this mechanism, important from a theoretical point
of view is that this mapping from a 4-ary DNA-subsequence to a 20-ary amino
acid sequence is non-trivial in the strict sense of definition 1.2.11 (page 36). This
mapping is defined by the tRNA molecules and reads, in natural organisms,
as given in table 1.2. It turns out that, in nature, the corresponding neutral
sets are typically a neighborhood with respect to genotypic variational topology.
Hence, some genotypic mutations are neutral, have no effect on the phenotype,
but change the genetic representation of that phenotype. The following is an
intriguing study of how nature exploits this phenomenon:

Example 1.4.1 *(Codon bias in HI viruses (Stephens & Waelbroeck 1999)).* Stephens
& Waelbroeck (1999) empirically analyze the codon bias and its effect in RNA
sequences of the HI virus. Several nucleotide triplets may encode the same amino
acid. For example, there are 9 triplets that are mapped to the amino acid Arginine. If Arginine is encoded by the triplet CGA, then the chance that a single
point mutation within the triplet is neutral (does not chance the encoded amino
acid) is $4/9$. In contrast, if it is encoded by AGA, then this *neutral degree* (definition 1.2.9, page 34) is $2/9$. Now, codon bias means that, although there exist
several codons that code for the same amino acid (which form a neutral set),
HIV sequences exhibit a preference on which codon is used to code for a specific
amino acid. More precisely, at some places of the sequence codons are preferred
that are "in the center of this neutral set" (with high neutral degree) and at
other places codons are biased to be "on the edge of this neutral set" (with
low neutral degree). It is clear that these two cases induce different exploration
densities; the prior case means low mutability whereas the latter means high mutability. Stephens and Waelbroeck go further by giving an explanation for these

---

coding sequence is copied onto a messenger RNA (mRNA, a sequence of nucleotides similar
to the DNA but with Uracil instead of Thymine). In Eukaryotes, when escaping the nucleus,
the mRNA is further processed and spliced (e.g., so-called introns are cut out). The following *translation* into a sequence of amino acids (poly-peptide chain) is probably the most
fascinating mechanism: There exist exactly 60 different transfer RNA (tRNA) molecules,
each of which has two parts. The "front" part is a matching key to a triplet of base pairs,
so-called *codons*, of which there exist $|\{A,U,G,C\}^3| = 64$ in number (4 of the 64 codons are
start or stop codons and need no translation). Depending on this codon key, the "back" part
of the tRNA is a matching key to one of 20 different amino acids. In this way, there is a
precise but non-injective match of a specific codon with a specific amino acid defined. (The
usual notation for all the tRNA with different front key but same back key is, e.g., $tRNA_1^{Leu}$,
$tRNA_2^{Leu}$ for the first and second tRNA molecule that binds to the amino acid Leucine.) The
translation process is as follows: The tRNA molecules are charged with matching amino acids
at their back part while the mRNA binds to the huge ribosome molecule (build itself out of
ribosomal RNA, rRNA). At the ribosome, matching tRNA molecules attach with their front
at base pair triplets of the mRNA and, since neighboring tRNA get so close (i.e., due to aw-
fully complicated mechanisms), the amino acids at their back parts join to the poly-peptide
chain.

two (marginal) exploration strategies: Loci with low mutability cause "more resistance to the potentially destructive effect of mutation", whereas loci with high mutability might induce a "change in a neutralization epitope which has come to be recognized by the [host's] immune system" (Stephens & Waelbroeck 1999).

We should mention another interesting implication of the triplet encoding concerning phenotypic variability: The insertion and deletion of a single nucleotide destroys the triplet rhythm and eventually changes the values of *all* succeeding codons. These kinds of mutations are called *frame shifts* and occur more frequently at certain repetitive sequences of the DNA. The interesting point is, again, that a single (non-correlated) mutation can lead to a highly correlated variation of *all* amino acids of the poly-peptide that gene codes for.

Another point concerning the mapping from DNA codons to amino acids is that the map-defining tRNA molecules are themselves encoded in the genome. Consider a mutation of a tRNA coding gene such that one of the tRNA molecules will induce another translation than usual. This would lead to a huge and highly correlated phenotypic variation since *all* DNA-codons of a given type will now be mapped to another amino acid. In today's organisms, such a mutation would definitely be lethal. At some time in early evolution though, the tRNA itself must have been evolved, i.e., there must have been mutations that have lead to the tRNA molecules that define the translation table 1.2. This phenomenon is often cited as an example for the "evolution of the genotype-phenotype mapping". We don't agree with such interpretations since the tRNA molecules are only one step in the whole mapping from genotype to phenotype and we think of the mapping *as a whole* as determined by natural laws. (We mentioned this issue before, in section 1.2.7 (page 39) "On evolving genetic representations", and will give a final point of view in section 1.4.4 when discussing so-called models of evolvable genotype-phenotype mappings.)

Protein folding.   The mapping from genotype to phenotype becomes more complex when considering the functionality of proteins instead of only their poly-peptide sequence. The functionality of a protein mainly depends on its physical 3-dimensional shape which determines whether the protein can bind to other molecules or not—many biological mechanisms really remind of keys matching to keyholes. Hence, much research is done on the question how a mere sequence of amino acids folds up to a complex tangled protein. Eventually, the law of

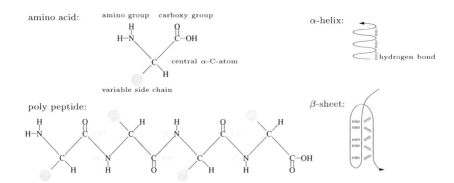

Table 1.3: Protein folding (from Knippers, 1997, chapter 1):
The peptide binding itself is a ridged surface (i.e., C_α-CO-N are in one (gray-shaded) plane). The N-C_α and C_α-C axes though are rotationally free. The interactions between the side chains determine the spatial structure of the protein.
Two rather regular structures are common, the α-helix and the β-sheet: α-helices arise from hydrogen bonds between the CO group of one amino acid and the NH group of the fourth next amino acid. Only parts of a protein are organized in such a helix (e.g., the amino acid Proline cannot have this hydrogen bond). β-sheets arise from hydrogen bonds between the CO and NH groups of distant (parallel or anti-parallel) subsequences of 5-10 amino acids. These sequences join together to form a sheet. Describing where a protein is organized as an α-helix or β-sheet means describing the secondary structure of the protein.

this *protein folding* are governed by chemical forces, mainly hydrogen bonds, between amino acids that become neighbored after folding. The whole problem though is too complex and high-dimensional to be simulated by computers for proteins of typical length (100-800 amino acids).

However, some regularities in the way proteins fold simplify the analysis: The α-helix and the β-sheet are a natural outcome of folding parts of the poly-peptide chain and thus common building blocks of proteins, see table 1.3. Hence, analyzing the structure of a protein may be decomposed as in the following definition: A protein's *primary structure* is its amino acid sequence. Knowing about a protein's *secondary structure* means knowing which parts of the amino acid sequence fold to α-helices or β-sheets. Finally, the *tertiary structure* describes how these α-helix and β-sheet building blocks are combined and wounded to form the protein's 3D shape.

Protein folding represents another part of the genotype-phenotype mapping, here mapping from the amino acid sequence to a rough description of the proteins functionality. Again, this mapping is highly neutral (non-injective) with extraordinary effects on phenotypic variability. Schuster (1996) and Fontana & Schuster (1998) show that the neutral set $[x]$ with respect to a given protein shape x (i.e., the set of all sequences that fold to the same shape) is *dense* in G. Here, dense means that for *any* sequence g in G, there exists at least one sequence g_x in the neutral set which is mutationally close to g. Mutational closeness can be measured by the number of possible mutations (given by the mutational topology on G) that are necessary to mutate from g_x to g. This shows how extended these neutral set are in G, that they figuratively percolate all through G. In our language this means that the space $\overline{[x]}$ of exploration distributions for a fixed phenotype x is very large and comprises are great variety of different exploration distributions $\sigma \in \overline{[x]}$.

In conclusion, the genotype-phenotype mapping in nature is highly non-trivial in the strict sense of definition 1.2.11 (page 36) and nature exploits the corresponding neutral sets as, e.g., shown in example 1.4.1. The way how gene sequences represent protein functionalities is essentially determined by (i) the laws of nature and (ii) the tRNA molecules. While the laws of nature were not subject to evolutionary adaptation, the tRNA molecules certainly were.

1.4.2   The operon and gene regulation: Structuring phenotypic variability

Considering the expression of genes as a phenotypic level, the operon is a paradigm for introducing correlations in phenotypic variability.

The previous section gave a first example of how a non-trivial phenotypic variability is induced by the mapping from DNA-codons to protein structures by virtue of the tRNA molecules and protein folding. Another remarkable mechanism to introduce structure in phenotypic variability is the operon, first described by Francois Jacob and Jacques Monod (Jacob & Monod 1961) for the lac operon of E. coli. Operon-like mechanisms will also be the basis of the computational model we will investigate in section 1.5.

The operon is a mechanism that correlates the expression of several genes in Prokaryotes. Typically, the operon consists of a DNA-subsequence in the fash-

ion ⟨promoter, operator region, several structural genes⟩ where each structural
gene is of the style ⟨start-codon, coding sequence, stop-codon⟩, i.e., it codes
for exactly one poly-peptide but lacks an own promoter. All these structural
genes are expressed at once if a polymerase binds to the operon's promoter and
if no other molecules prohibit the polymerase to proceed by attaching to the
operator region. Molecules that inhibit the expression of structural genes by
binding to the operator region are called *repressors*, and such mechanisms are
generally called *gene regulation*. Typically, a repressor indicates whether the
expression of the genes is necessary, e.g. for the metabolism, or not. But gene
regulation may become arbitrarily complex, reminding at complex logic dependencies: Some repressors bind to the operator region only in conjunction with an
another molecule, the *corepressor* (an "NOT A OR NOT B" conjunction: if either
the repressor OR the corepressor is missing, the genes are expressed). In other
cases, another molecule, the *inducer*, may bind to the repressor and prohibit
its binding to the operator region (an "NOT A OR B" conjunction: if either the
repressor is missing OR an inducer is present, the genes are expressed).

Example 1.4.2 *(The lac operon in E. coli).* The lac operon in E. coli bacterium
comprises three structural genes which code for enzymes that are necessary to
metabolize lactose in the cell's environment. If there is *no* lactose in the cellular
environment, a repressor protein (which is constantly expressed by the *I gene*)
binds the the operon's operator region and prevents the expression of the three
genes. If lactose is present, the lactose binds to the repressor protein such that
it looses its affinity to the operator region; the operator region is freed and the
three genes are expressed.

Eventually, any combination of logic kinds of gene regulation mechanisms becomes possible if accounting that repressors, corepressors, and inducers may
themselves be encoded on genes or even on structural genes of other operons—
mechanisms that are called *gene interaction*.

Again, what does that mean for our discussion of genotype and phenotype variability? The point here is that gene regulation introduces high correlations
between the expression of genes. For example, as a partial genotype-phenotype
mapping consider the relation between the DNA and the production rates of
proteins in a given cellular molecular environment. (In the previous section, we
captured these production rates by the operator Π.) If variability on the DNA
is rather simply structured, hypercube-like, without correlations between single
1st-type mutations, there are high and complex structured correlations in the
variability of the production rates: One single mutation on the DNA may change

the promoter of an operon such that all the structural genes in the operon are not expressed; the production rates of the corresponding proteins decrease *in correlation*. Similarly, one single mutation in a gene that codes for a repressor, corepressor, or inducer may effect, via the recursive dependencies of gene regulation, the expression of many genes; the production rates of the corresponding proteins all vary in a correlated way, where the correlations are determined by the kind of gene interactions.

Again, it is important to note that these mechanisms of gene regulation, which determine the correlatedness of protein production, are themselves encoded in the genome in terms of operator and promoter regions. Hence, these interactions themselves evolved and again the interpretation is tempting that the genotype-phenotype mapping (in this case the part of the GP-map the maps the 4-ary DNA sequence on the rates of protein production) is evolvable. As in the previous section we regard this a misleading interpretation since the GP-map as a hole is fixed and determined by the laws of nature (in this case the molecular laws that determine the binding of enzymes and repressors, etc. to the DNA).

In section 1.3.1 (page 49) we already presented an intriguing example of complex phenotypic variability due to gene regulation: the eyeless gene of Drosophila Melanogaster. See figure 1.2 for these phenotypic variations induced by a mutation of a single gene. Another example is the following:

Example 1.4.3 *(Mice's brain).* Chenn & Walsh (2002) demonstrated that the enforced expression of a single protein (β-catenin) during brain development of a mouse enlarges the brain hugely. Figure 1.3 shows this enlargement and the human-like folding of the tissue. Again, there exists a single control mechanisms to produce this highly correlated phenotypic variation.

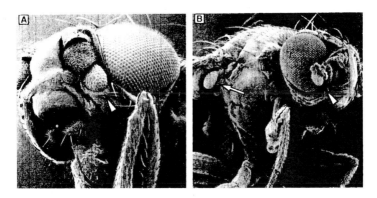

Figure 1.2: The arrows point a functionally complete eye modules that grew at various places due to a single gene mutation of Drosophila Melanogaster; see example 1.3.1 (page 49). Reprinted from (Halder, Callaerts, & Gehring 1995) with permission of the authors.

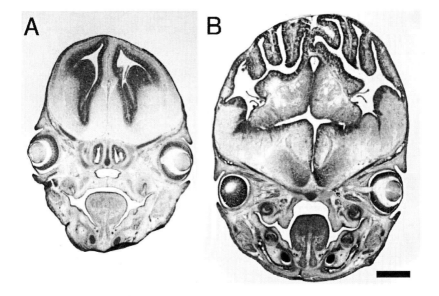

Figure 1.3: See example 1.4.3. Reprinted from (Chenn & Walsh 2002) with permission of the authors.

1.4.3   Flowers and Lindenmayer systems: An artificial model for a genotype-phenotype mapping

Grammar-like genotype-phenotype mappings realize complex, correlated vari-ability in artificial evolutionary systems—similar to the operon—and in-duce highly non-trivial neutrality.

After discussing genotypic and phenotypic variability in natural systems we want to present a more abstract example: Prusinkiewicz & Hanan (1989) proposed an encoding for plant-like structures which they use to produce incredible realistic computer generated pictures of flowers and trees (Prusinkiewicz & Lindenmayer 1990). The reason we refer to them is because their encoding is an excellent ex-ample for recursive, grammar-like, highly *non-trivial* genotype-phenotype map-pings.

The version of the encoding we describe here is based on sequences of the alpha-bet {F,+,-,&,^, \,/,[,],.}. The meanings of these symbols are described in table 1.5. For example, the sequence ⟨FF[+F][-F]⟩ represents a plant for which the stem grows two units upward before it branches in two arms, one to the right, the other to the left, each of which has one unit length and a leave attached at the end.

The *L-systems* proposed by Lindenmayer are similar to grammars that produce such sequences. For example, the L-system

start-symbol=X, X→F[+X]F[-X]+X, F→FF

produces, after some recursions of the replacement rules, the sequences

⟨X⟩

⟨FF[+X]FF[-X]+X⟩

⟨FFFF[+FF[+X]FF[-X]+X]FFFF[-FF[+X]FF[-X]+X]+FF[+X]FF[-X]+X⟩

etc.

Table 1.4 demonstrates the implications of this encoding. The two plants on the left are examples taken from (Prusinkiewicz & Lindenmayer 1990). To the right of these "originals" you find three different variations of both plants. Each variation was produced by a *single* symbol mutation in the genotype.

Think of the phenotype, i.e., the plant structure, as a long list of coordinates that describe the position and orientation of each element of the plant (similar to how a computer would store a 2D vector graphics). If we now assume a single decor-

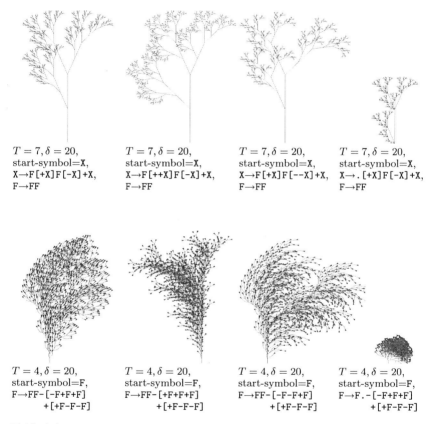

Table 1.4: Two examples for 2D plant structures and their phenotypic variability induced by *single* symbol mutations. T denotes the number of iterations of the production rules to generate the phenotype; δ is the rotation angle as described in table 1.5.

F	attach a unit length stem
+,-,&^,\,/	rotations of the "local coordinate system" (that apply on following attached units): four rotation of δ degree in all four directions away from the stem (+,-,&,^) and two rotations of $\pm\delta$ degree around the axis of the stem (\,/).
[	a branching (instantiation of a new local coordinate system)
]	the end of that branch: attaches a leave and proceeds with the previous branch (return to the previous local coordinate system); also if the very last symbol of the total sequence is not], another leave is attached
.	does nothing

Table 1.5: Description of the plant grammars symbols, cf. (Prusinkiewicz & Hanan 1989).

related mutation distribution on the genotype (where each symbol is mutated with probability α) what does the variation distribution on phenotype space look like? What the pictures demonstrate is that there would be "large-scale" correlations between the coordinates that describe the plant; in the examples, since there are no decoupled subtrees, even all elements of the plant would be correlated w.r.t. this phenotypic variation distribution.

The computational model in the next section will be based on a similar grammar-like genetic encoding. One of the experiments will even consider the evolution of artificial plants. In principle, the application of grammar-like genotype-phenotype mappings in evolutionary computation is not new. However, there exists no approach in the literature that exploits σ-evolution, i.e., allows for neutral mutations that enable transitions between different phenotypic variation distributions. We will do so in our computational model.

1.4.4   Altenberg's evolvable genotype-phenotype map

Ultimately, discussing Lee Altenberg's model of the evolution of the genotype-phenotype mapping pinpoints our point of view on the subject.

Altenberg (1995) proposed a model of the *evolution of the genotype-phenotype map* which comprises three main elements:

- bit strings $x \in \{0,1\}^n$ of length n,

- a $n \times f$-matrix $M \in \{0,1\}^{n \times f}$ of 0s and 1s,

- and a fixed but randomly chosen function $\Phi : \{0,1\}^n \times \{1, .., f\} \times \{0,1\}^n \to [0,1]$.

He calls the strings x "genotypes" and M and Φ allow to calculate a fitness for each x as follows: For each x we calculate f different fitness components ϕ_j, $j = 1..f$, by first calculating the bit-wise AND between the x and the j-th column of M, $\bar{x}_j = x$ AND $(M._j)$, and then defining $\phi_j(x) = \Phi(\bar{x}_j, j, (M._j))$. The total fitness of x is simply the average $\frac{1}{f} \sum_j \phi_j(x)$ of these fitness components. This means that the j-th column of M determines on which bits the j-th fitness component generally depends, whereas Φ determines *how* it depends on these bits. This model is thus a generalization of the NK-landscape models (Kauffman 1989) where N fitness components always depend on K (randomly chosen) bits for string of fixed length N.

Altenberg interprets M as the genotype-phenotype mapping. If there are many 1s in a column, the respective fitness component depends on many genes, i.e., *polygeny* is high. If there are many 1s in a row, the respective gene influences many fitness components, i.e., *pleiotropy* is high.

The model for the evolution of the genotype-phenotype mapping considers only one individual x and is an alternation between a phase of hill climbing by adapting x without changing M or Φ and a phase of genome growth where n is increased by 1 and the new gene in x and the new row in M are chosen randomly. The hill climbing phase lasts until a local optimum is found. During genome growth the addition of a new gene is accepted only when it results in no decrease of total fitness. Otherwise the gene addition is discarded and "another gene" is added (i.e., the new gene value and the new row of M are again chosen randomly). The genome growth phase lasts until a new gene has been added without fitness decrease.

How should one interpret this model?   Altenberg demonstrates very interesting effects with this model, e.g., that during evolution pleiotropy (as defined above) is decreased—an effect that he calls "constructional selection for low pleiotropy." But let us first discuss how one should interpret the model itself. Since Altenberg calls M the GP-map and M evolves by continuous addition of rows during genome growth, he considers his model as an example for the evolution of the GP-map. But is M really the mapping from genotype to phenotype?

What are actually the genotype and phenotype spaces in his model?

Let us first consider the space of all strings of any length as genotype space,

$$G = \bigcup_{n=1}^{n_{max}} \{0,1\}^n \ .$$

And let us consider the tuple of vectors $(\bar{x}_1, .., \bar{x}_n)$ with $\bar{x}_j = x$ AND $(M_{.j})$ as the phenotype. In this interpretation the fitness function from phenotype to a real number is fixed and well-defined via Φ and, indeed, M specifies the mapping from genotype to phenotype. But the question becomes: When you add a single gene to the genome x, namely add one bit to the bit-string, why should this gene have *random* effects on the phenotype, i.e., why should the additional row in M (which is supposed to correspond to the pleiotropic effects of the new gene) be *random*? In nature, when you add a specific DNA sequence to a genome, the effects are not random but in some way (be it stochastically) determined by the laws of nature.

For that reason one should interpret the new row of M not as a character of the GP-map but as a character of the new gene. But all characters of a gene are determined by the sequence of the gene. Hence, the new row of M is actually an abstract way of specifying a new gene in terms of its pleiotropic effects instead of by its sequence—and thus M is part of the genotype! In that interpretation the genotype space comprises the matrices,

$$G = \bigcup_{n=1}^{n_{max}} \{0,1\}^n \times \{0,1\}^{n \times f} \ .$$

The genotype-phenotype mapping $G \rightarrow P$, $(x, M) \mapsto (\bar{x}_1, .., \bar{x}_n)$, is well-defined, fixed, and not subject to evolution. This interpretation challenges none of the results presented by Altenberg—the genotype (x, M) evolves, via constructional selection, such that pleiotropy between genes is decreased. But it challenges the statement that this is due to an evolution of the GP-map.

This discussion should be transferred to the scenarios we discussed earlier. E.g., the tRNA molecule could be compared with M: it characterizes the effects of genes. But the evolution of the tRNA molecules does not imply an evolution of the GP-map; the tRNA is itself specified by the genome. The same holds for any kind of mechanisms that determine the phenotypic effects of genes but are themselves determined by the genome as a whole (e.g., gene regulation mechanisms which are determined by the sequences of promoters, inhibitors, etc.).

1.5   A computational model

The model we introduce and discuss in this section is meant to demonstrate
the evolution of genetic representations in the case of a non-trivial GP-map.
The main point will be a specific genotype-phenotype mapping that we introduce as an abstraction and simplification of ontogenetic mechanisms. Although
the model is rather simple, it offers a very complex variability of exploration
distributions that goes beyond existing models.

1.5.1   The genotype and genotype-phenotype mapping

*An abstract model of ontogenetic development mimics the basic operon-like
mechanism to induce complex phenotypic variability.*

Let us consider the development of an organism as an interaction of its state Ψ
(describing, e.g., its cellular configuration) and a genetic system Π, neglecting
environmental interactions. Development begins with an initial organism in
state $\Psi^{(0)}$, the "egg cell", which is also inherited and which we model as part
of the genotype. Then, by interaction with the genetic system, the organism
develops through time, $\Psi^{(1)}, \Psi^{(2)}, .. \in P$, where P is the space of all possible
organism states. Hence, the genetic system may be formalized as an operator
$\Pi : P \to P$ modifying organism states such that $\Psi^{(t)} = \Pi^t \Psi^{(0)}$. We make a
specific assumption about this operator: We assume that Π comprises a whole
sequence of operators, $\Pi = \langle \pi_1, \pi_2, .., \pi_r \rangle$, each $\pi_i : P \to P$. A single operator π_i
(also called production rule) is meant to represent a transcription module, i.e.,
a single gene or an operon. Based on these ideas we define the general concept
of a genotype and a genotype-phenotype mapping for our model:

*A genotype consists of an initial organism $\Psi^{(0)} \in P$ and a sequence $\Pi =
\langle \pi_1, \pi_2, .., \pi_r \rangle$, $\pi_i : P \to P$ of operators. A genotype-phenotype mapping ϕ de-
velops the final phenotype $\Psi^{(T)}$ by recursively applying all operators on $\Psi^{(0)}$.*

This definition is somewhat incomplete because it does not explain the stopping
time T of development and in which order operators are applied. We keep to
the simplest options: We apply the operators in sequential order and will fix T
to some chosen value.

For the experiments we need to define how we represent an organism state Ψ and how operators are applied. We represent an organism by a sequence of symbols $\langle\psi_1, \psi_2, ..\rangle$, $\psi_i \in \mathcal{A}$. Each symbol may be interpreted, e.g., as the state of a cell; we choose the sequence representation as the simplest spatially organized assembly of such states. Operators are represented as replacement rules $\langle a_0{:}a_1, a_2, ..\rangle$, $a_i \in \mathcal{A}$, that apply on the organism by replacing all occurrences of a symbol a_0 by the sequence $\langle a_1, a_2, ..\rangle$. If the sequence $\langle a_1, a_2, ..\rangle$ has length greater than 1, the organism is growing; if it has length 0, the organism shrinks. Calling a_0 promoter and $\langle a_1, a_2, ..\rangle$ the structural genes gives the analogy to operons in natural genetic systems. For example, if the initial organism is given by $\Psi^{(0)}{=}\langle a\rangle$ and the genetic system is $\Pi{=}\langle$ $\langle a{:}ab\rangle, \langle a{:}cd\rangle, \langle b{:}adc\rangle$ $\rangle$, then the organism grows as: $\Psi^{(0)}{=}\langle a\rangle$, $\Psi^{(1)}{=}\langle cdadc\rangle$, $\Psi^{(2)}{=}\langle cdcdadcdc\rangle$, etc.

The general idea is that these operators are basic mechanisms which introduce correlating effects between phenotypic traits. Riedl (1977) already claimed that the essence of the operon is to introduce correlations between formerly independent genes in order to adopt the *functional* dependence between the genes and their phenotypic effects and thereby increase the probability of successful variations.

The proposed model is very similar to the models by Kitano (1990), Gruau (1995), Lucas (1995), and Sendhoff & Kreutz (1998), who use grammar-encodings to represent neural networks. It is also comparable to new approaches to evolve complex structures by means of so-called *symbiotic composition* (Hornby & Pollack 2001a; Hornby & Pollack 2001b; Watson & Pollack 2002).

The crucial novelty in our model are specific *2nd-type mutations*. These allow for genetic variations that explore the neutral sets which are typical for any grammar-like encoding. Without these neutral variations, self-adaptation of genetic representations and of exploration distributions is not possible.

1.5.2   2nd-type mutations for the variability of exploration

A new type of structural mutations allows for reorganizations of the genetic representations and exploration of the respective neutral sets.

Consider the three genotypes given in the first column,

genotype	phenotype	phenotypic neighbors
$\Psi^{(0)}=\langle\text{a}\rangle,\,\Pi=\langle\;\langle\text{a:bcbc}\rangle\;\rangle$	$\langle\text{bcbc}\rangle$	$\langle*\rangle,\,\langle*\text{cbc}\rangle,\,\langle\text{b}*\text{bc}\rangle,\,\langle\text{bc}*\text{c}\rangle,\,\langle\text{bcb}*\rangle$
$\Psi^{(0)}=\langle\text{a}\rangle,\,\Pi=\langle\;\langle\text{a:dd}\rangle,\langle\text{d:bc}\rangle\;\rangle$	$\langle\text{bcbc}\rangle$	$\langle*\rangle,\,\langle*\text{bc}\rangle,\,\langle\text{bc}*\rangle,\,\langle*\text{c}*\text{c}\rangle,\,\langle\text{b}*\text{b}*\rangle$
$\Psi^{(0)}=\langle\text{bcbc}\rangle,\,\Pi=\langle\;\;\;\rangle$	$\langle\text{bcbc}\rangle$	$\langle*\text{cbc}\rangle,\,\langle\text{b}*\text{bc}\rangle,\,\langle\text{bc}*\text{c}\rangle,\,\langle\text{bcb}*\rangle$

All three genotypes have, after developing for at least two time steps, the same
phenotype $\Psi^{(t)}=\langle\text{bcbc}\rangle$, $t \geq 2$. The third genotype resembles what one would
call a direct encoding, where the phenotype is directly inherited as $\Psi^{(0)}$. Assume
that, during mutation, all symbols, except for the promoters, mutate with fixed,
small probability. By considering all one-point mutations of the three genotypes,
we get the phenotypic neighbors of $\langle\text{bcbc}\rangle$ as given in the third column of the
table, where a star $*$ indicates the mutated random symbol. These neighborhoods
differ significantly; we have an example for a neutral set such that phenotypic
variational topology along this set varies. The neighborhood induced by the
second genotype is particularly interesting: A mutation on the first operator
leads to neighbors $\langle*\text{bc}\rangle$, $\langle\text{bc}*\rangle$ where the whole "module" bc is replaced by
another symbol. On the other hand, a mutation on the second operator leads to
neighbors $\langle*\text{c}*\text{c}\rangle$ and $\langle\text{b}*\text{b}*\rangle$ such that the two stars in each neighbor represent
the *same* random symbol; the module bc itself is mutated and duplicated. Thus,
there exist high correlations in the exploration distribution between symbols
within one module and the corresponding symbols of different modules. This
is a first example of how correlations and the hierarchical kind of genotype-
phenotype mapping leads to a notion of modularity.

In order to enable a variability of genetic representations within such a neutral set
we need to allow for mutational transitions between phenotypically equivalent
genotypes. A transition from the 1st to the 3rd genotype requires a genetic
mutation that applies the operator $\langle\text{a:bcbc}\rangle$ on the egg cell $\langle\text{a}\rangle$ and deletes it
thereafter. Both, the application of an operator on some sequence (be it the
egg cell or another operator's) and the deletion of operators will be mutations
we provide in the computational model. The transition from the 2nd to the
1st genotype is similar: The 2nd operator $\langle\text{d:ab}\rangle$ is applied on the sequence of
the first operator $\langle\text{a:dd}\rangle$ and deleted thereafter. But we must also account for
the inverse of these transitions. A transition from the 3rd genotype to the 1st is
possible if a new operator is created by randomly extracting a subsequence (here
bcbc from the egg cell) and encoding it in a new operator (here $\langle\text{a:bcbc}\rangle$). The
original subsequence is then replaced by the promoter. Similarly, a transition
from the 1st to the 2nd genotype occurs when the subsequence bc is extracted

• First type mutations are ordinary symbol mutations that occur in every sequence (i.e., promoter or rhs of an operator) of the genotype; namely symbol replacement, symbol duplication, and symbol deletion, which occur with equal probabilities. The mutation frequency for every sequence is Poisson distributed with the mean number of mutations given by ($\alpha \cdot$ sequence-length), where α is a global mutation rate parameter.

• The second type mutations aim at a neutral restructuring of the genetic system. A 2nd-type mutation occurs by randomly choosing an operator π and a sequence p from the genome, followed by one of the following operations:

  – application of an operator π on a sequence p,
  – inverse application of an operator π on a sequence p; this means that all matches of the operator's rhs sequence with subsequences of p are replaced in p by the operator's promoter,
  – deletion of an operator π, only if it was never applied during ontogenesis,
  – application of an operator π on *all* sequences of the genotype followed by deletion of this operator,
  – generation of a new operator ν by extracting a random subsequence of stochastic length $2 + \text{Poisson}(1)$ from a sequence p and encoding it in an operator with random promoter. The new operator ν is inserted in the genome behind the sequence p, followed by the inverse application of ν on p.

All these mutations occur with equal probabilities. The total number of second type mutations for a genotype is Poisson distributed with mean β. The second type mutations are not necessarily neutral but they are neutral with sufficient probability to enable an exploration of neutral sets.

• A genotype is mutated by first applying second type mutations and thereafter first type mutations, the frequencies of which are given by β and α, respectively.

Table 1.6: The mutation operators in our model.

from the operator $\langle$a:bcbc$\rangle$ and encoded in a new operator $\langle$d:bc$\rangle$.

Basically, the mutations we will provide are the generation of new operators by extraction of subsequences (*deflation*), and the application and deletion of existing operators (*inflation*). Technical details can be found in table 1.6; the main point of these mutation operators though is not their details but that they in principle enable a transition between phenotypically equivalent representations in our encoding.

1.5.3 First experiment: Neutral σ-evolution

Genetic representations reorganize themselves in favor of short genome length and modular phenotypic variability.

λ	100	(offspring) population size
μ	30	(selected) parent population size
α	.03	frequency of symbol replacement mutations
	.0	frequency of symbol insertion and deletion mutations
β	.1	frequency of all second type mutations
$\mathcal{A}$	a,..,h	symbol alphabet
T	1	stopping time of development
	2000	number of samples to analyze the exploration distributions

Table 1.7: The parameters we use in the first experiment

Specification. The first experiment is designed to demonstrate the dynamics of σ-evolution on a neutral set. We initialize a population of $\lambda = 100$ genotypes on a single point in the neutral set and investigate how evolution explores the rest of the neutral set. The neutral set comprises all possible encodings of the fixed, self-similar phenotype ⟨abcdeabcdeabcdeabcdeabcde⟩, i.e. 5× the sequence abcde. We choose the direct encoding, without operators and $\Psi^{(0)}$ equal to the phenotype, to initialize the population. The mutations we introduced in the last section allow for the development of more complex genetic representations.

The mutations, the genotype-phenotype mapping, and the initialization have been specified so far. The selection we utilize is the most simplest one could think of: Only "correct" phenotypes, equal to the one given by the neutral set that we investigate, are selected. The algorithmic implementation is also very simple; it is the so-called (μ, λ)-selection (Schwefel 1995). We randomly select $\mu = 30$ parents out of the population (it never occurred that we had less than μ correct phenotypes in the population), generate $\lambda = 100$ clones of them (by roulette-wheel selection), and mutate this offspring. We simulated no crossover. The few parameters of the simulation are summarized in table 1.7.

Measures. We utilized several measures to analyze the dynamics of σ-evolution. Three of them are features of the phenotypic exploration distribution $\Xi\sigma$ and are calculated from a finite size sample (of size 2000) of the distribution for each individual. First, this is the neutral degree

$$n = (\Xi\sigma)(x) \,,$$

as we defined it in 1.2.9 (page 34), where x is the parent's phenotype, in our case
the "correct phenotype". Second, we calculate the mutual information between
phenotypic variables in the exploration distribution $\Xi\sigma$,

$$I_{ij} = \sum_{a,b\in\mathcal{A}} \Xi\sigma_{ij}(a,b) \ln \frac{\Xi\sigma_{ij}(a,b)}{\Xi\sigma_i(a)\,\Xi\sigma_j(b)} \; ,$$

where $\Xi\sigma_i$ denotes the marginal distribution of $\Xi\sigma$ in the i-th variable, and
$\Xi\sigma_{ij}$ the marginal in both, the i-th and j-th variable. We display the mutual
information by normalizing with the marginal entropies,

$$I'_{ij} = \frac{2\,I_{ij}}{H_i + H_j} \; , \quad H_i = \sum_{a\in\mathcal{A}} \Xi\sigma_i(a) \ln \Xi\sigma_i(a) \; ,$$

and drawing I'_{ij} as a gray-shade matrix where black corresponds to 0 and white
to 1. The third measure, the *modular degree*, we specifically introduce for our
scenario: It is the sum of probabilities that, when a variation occurs at position i
of the phenotype, the same variation occurs also at a position $[(i+k\cdot 5) \mod 25]$,
summed over $k = 1..4$. The positions $[(i + k \cdot 5) \mod 25]$ have distance 5, 10,
15, or 20 from the position i and represent corresponding variables of different
modules. If the modular degree is high, then mutations within one module are
likely to occur in the same way also in other modules—it is thus a measure of
the self-similarity of variability.

Two further measures give simple information about the genetic representation.
The *genome length* is the sum of lengths of the egg cell $\Psi^{(0)}$ and all operators
π_i. The *operator usage* is the number of operators that have been applied during
ontogenesis.

Results.   Figures 1.4, 1.5, and 1.6 display the result of the simulation. From
figure 1.4 one gets an idea on what kind of genotypes are explored during σ-
evolution. It displays the genotypes and mutual information matrices of exploration for a single randomly picked individual in selected generations. All
genotypes represent of course the same phenotype but evolution has managed to
generate new operators by extracting and encoding subsequences. The modular
encoding of these subsequences leads to the correlations between certain phenotypic variables as displayed by the mutual information matrices. For example,
the representation found in generation 21 has extracted a subsequence of length
3 and the mutual information matrix exhibits the correlations between those

gene-ration	genotype	I'_{ij}	gene-ration	genotype	I'_{ij}
0	⟨abcdeabcdeabcde... ...abcdeabcde⟩		215	⟨fffff⟩ ⟨f:he⟩ ⟨h:abcd⟩	
21	⟨abchbchbchbchbcde⟩ ⟨h:dea⟩		220	⟨hehehehehe⟩ ⟨h:abcd⟩	
125	⟨abfhbfhbcde⟩ ⟨f:chbc⟩ ⟨h:dea⟩		225	⟨eef⟩ ⟨e:ff⟩ ⟨f:abcde⟩	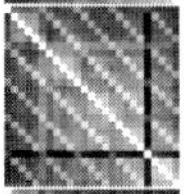
145	⟨adadadadad⟩ ⟨d:bcde⟩		255	⟨f⟩ ⟨f:fffff⟩ ⟨f:abcde⟩	
150	⟨fffff⟩ ⟨f:ace⟩ ⟨c:bcd⟩		270	⟨fffff⟩ ⟨f:abcde⟩	
200	⟨fffff⟩ ⟨f:abcde⟩				

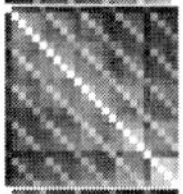

Figure 1.4: Different genotypes encode the same phenotype of length 25 but with different exploration distributions. The 25×25-matrix I'_{ij} comprises the normalized mutual informations between two phenotypic variables in the phenotypic exploration distribution $\Xi\sigma$. White means 1 and black means 0. The diagonals are white because $I_{ii} = H_i$. The regular strips in the matrix exhibit the correlations between symbols (typically they have distance 5 in the phenotype sequence!). The first symbols of the phenotype sequence correspond to the upper left corner of the matrix, the last symbols to the lower right. E.g., the matrix in generation 225 shows that the last module is encoded differently than the others.

phenotypic variables that stem from the same variable on the first operator. In the next genetic reformation, at generation 125, a second operator is introduced leading to more correlations. Further reformations occur until, in generation 200, a genetic representation is found that one might consider optimal. It is the representation of minimal length 11. In the sequel, other genetic representations are found, but all similar to the optimal one.

Since the evolution occurs only on a neutral set, there is no need to display

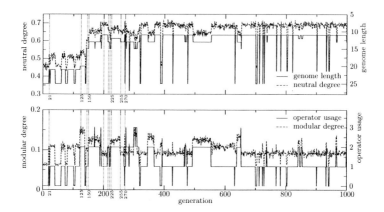

Figure 1.5: Features of the phenotypic exploration distribution for the same single individual as for figure 1.4. Notice that the scale of the genome length in the upper graph is inverted. The correlations between neutral degree and genome length and between modular degree and operator usage are evident. The use of appropriate operators coincides with a more structured, modular exploration distribution, a decrease of genome length and thus higher neutral degree. The vertical dotted lines correspond to the generations depicted in figure 1.4.

a fitness curve. Instead, figures 1.5 and 1.6 directly illustrate the evolution of phenotypic exploration distributions by displaying features of $\Xi\sigma$. Figure 1.5 shows the curves for the same individual as in figure 1.4. Notice that the scaling of the genome length has been inverted to emphasize the strong relation between genome length and neutral degree. Evolution starts with a representation of length 25 and rather low neutral degree of 0.45. Until about generation 200, the genome length decreases significantly, down to length 11, and the neutral degree increases correspondingly to 0.7. The decrease of genome length is achieved by using operators which, in turn, induce these special kind of correlations measured by the modular degree. After generation 200, all four measures fluctuate in high correlation. Figure 1.6 displays the same four measures, but now averaged over the whole population of $\lambda = 100$ offspring. The upper two curves show the strong pressure towards small genome length and high neutral degree. And the lower two curves show the concise impact of operator usage on the modularity of the exploration distribution.

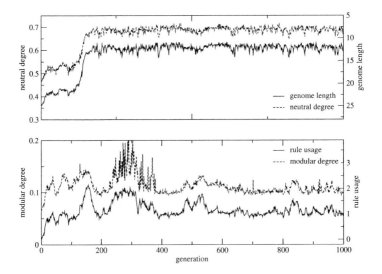

Figure 1.6: Features of the phenotypic exploration distribution averaged over the population. Their correlations and the dynamics toward higher neutral degree are apparent.

Validation of theory. In summary, what we find is that σ-evolution, here exclusively on a neutral set, indeed realizes an exploration of the neutral set and a great variability of phenotypic exploration distributions with respect to the three features we tracked (neutral degree, modular degree, and mutual information). The dynamics of σ-evolution are such that the neutral degree increases together with certain correlations and modular dependencies. We have to check that this is consistent with the theoretical findings of the previous section.

The selection mechanism was such that "correct" phenotype are selected with equal probability. This means that for each generation t we may choose a selection

$$\tilde{f}^{(t)}(p) = \begin{cases} \frac{1}{c^{(t)}} & \text{if } p = p_{\text{correct}} \\ 0 & \text{if } p \neq p_{\text{correct}} . \end{cases}$$

where $p_{\text{correct}} \in P$ is the correct phenotype and $c^{(t)}$ is the number of correct phenotypes in this generation. The σ-quality of an exploration distribution σ is

$$\langle \tilde{f}^{(t)} , \sigma \rangle = \sum_{p \in P} f^{(t)}(p) \,\Xi\sigma(p) = \frac{1}{c^{(t)}} \,\Xi\sigma(p_{\text{correct}}) = \frac{n}{c^{(t)}} \,, \qquad (1.14)$$

where $p_{\text{correct}} \in P$ is the correct phenotype and n the neutral degree of σ. Thus, at each time, σ-quality is proportional to the neutral degree.

The neutral degree is actually a measure of mutational robustness, i.e., the chance that mutations do not destroy a phenotype. In correspondence to Wagner (1996), our experiment demonstrates the evolution of evolutionary plasticity towards mutation robustness under the condition that evolution is not dominated by innovative progress.

1.5.4   Second experiment: σ-evolution for phenotypic innovation

Genetic representations develop in favor of highly correlated phenotypic variability allowing for simultaneous phenotypic innovations.

The second experiment demonstrates the benefit of 2nd-type mutations for phenotypic adaptation. The scenario remains the same as for the first experiment except for the initialization and selection. We initialize each individual as $\Psi^{(0)} = \langle \mathsf{a} \rangle$ and no operators. The μ individuals that are closest to the correct phenotype are selected ((μ, λ)-selection with respect to Hamming distance). We investigate four different cases: We study the dynamics with and without 2nd-type mutations ($\beta = 0.1$ and $\beta = 0$) once for moderate mutation rate ($\alpha = 0.03$) and once with higher mutation rate ($\alpha = 0.06$) w.r.t. symbol replacement, insertion, and deletion. All remaining parameters are the same as for the first experiment. We perform 10 independent trials for each case, see figure 1.7.

The two graphs on the left refer to the case of moderate mutation rate $\alpha = 0.03$, where moderate means that $\alpha < \frac{1}{25}$ (recall the length 25 of the correct phenotype). The upper graph displays the fitness curves with 2nd-type mutations enabled, $\beta = 0.01$, whereas for the lower graph $\beta = 0$ and no 2nd-type mutations occur. Without 2nd-type mutations, neutral sets are not explored, no operators are created and the encoding remains a direct one, as it was initialized. In both cases the optimal phenotype is found, but the characteristics of evolution are

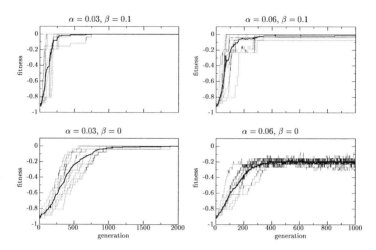

Figure 1.7: The fitness depicted here is the negative percentage of symbols of a phenotype that do not match with the correct phenotype. The four graphs display the evolution with and without 2nd-type mutations ($\beta = 0.1$ and $\beta = 0$) and with low and high mutation rate ($\alpha = 0.03$ and $\alpha = 0.06$). Each graph contains the dotted curves of 10 independent trials and their average. Notice that the time scales are different for different mutation rates α.

very different. For the direct encoding, innovation occurs as successive adaptation of single symbols what becomes apparent by the successive small steps in the fitness curve. Instead, with 2nd-type mutations enabled, a few small steps of innovation occur at the beginning; then huge steps of innovation occur when 2nd-type mutations have generated operators, changed the exploration distribution, and enabled a modular and self-similar growth of the phenotype; and further smaller steps of adaptation follow until the optimum is reached. These last steps can be interpreted as symbol-wise adaptation of the tail and head of the phenotype (which might not be encoded modularly from the beginning, see below). Evolution with 2nd-type mutations is much faster due to the better adapted (modular, self-similar) exploration distributions. Here, the selective pressure in σ-evolution is not only the neutral degree but the symmetric structure of the fitness distribution, which is *not* concentrated only at the optimum as for the first experiment.

The two graphs on the right display the case of high mutation rate $\alpha = 0.06$.

The direct encoding of a phenotype of length 25 is not stable and the curves for
the direct encoding case (lower plot) do not converge towards the optimal phenotype but fluctuate around 20% non-optimal symbols. Instead, when 2nd-type
mutations are enabled, $\beta = 0.01$, evolution arranges to find a stable encoding of
the optimal phenotype by using operators and decreasing description length.

We observe an important phenomenon of σ-evolution in our model. The neutral
genetic variations supposed to explore the variety of phenotypically equivalent
encodings also change the description length. However, in the case of high mutation rates, encodings of long description lengths are not stable. Often though, a
transition between different compact encodings requires first a neutral inflation
of the genome (dissolving and removing operators and approaching a more direct
encoding) before deflating it again. The intermediate state of long description
length, although it is phenotypically equivalent, is a barrier for σ-evolution because in the case of high mutation rate its σ-quality is very low. These barriers
inhibit the exploration of the whole neutral set and cause local optima with respect to σ-quality. In some cases, this also inhibits the transition toward the
optimal phenotype.

Four of the 10 runs (for $\alpha = 0.06$ and $\beta = 0.01$) exemplify this phenomenon.
Two runs did not at all find the optimal phenotype, the population centered
around the following genotypes:

genotype	phenotype
$\langle$abgdeacccc$\rangle$, $\langle$c:bc$\rangle$, $\langle$c:cdea$\rangle$	$\langle$abgde abcde abcde abcde abcdea$\rangle$
$\langle$abcddddae$\rangle$, $\langle$d:deabc$\rangle$	$\langle$abcde abcde abcde abcde abcae$\rangle$

The first genotype maneuvered itself into a dead-end in several respects. The g
(3rd symbol of $\Psi^{(0)}$) cannot be replaced by a c since c is used as a promoter. The
only chance would be to dissolve the two operators by inflation and thereafter
create new operators with other promoters than c. This however would require
to trespass the barrier of long description length. Further, the last symbol of the
phenotype (the a) cannot be removed because it is needed in the 2nd operator
as part of the module bcdea. The genetic system implemented the module bcdea
instead of abcde. The situation is similar for the second genome. The symbol a
(second to last) in $\Psi^{(0)}$ cannot be replaced by a d because d is used as a promoter.
Again, only an inflation would enable an advantageous genetic reformation.

Two further runs did not realize the "optimal encoding" of length 11, the population centered around the following genomes:

genotype	phenotype
⟨abcffffde⟩, ⟨f:deabc⟩	⟨abcde abcde abcde abcde abcde⟩
⟨accccd⟩, ⟨c:da⟩, ⟨d:bcde⟩	⟨abcde abcde abcde abcde abcde⟩

Both of them realize this sort of unfortunate modularization; the first implements modules deabc, the second bcdea. This kind of modularization may stem from earlier epochs of evolution where subsequences are extracted from the middle. Consequently, the head and tail of the final phenotype are not encoded by operators. A restructuring in favor of shorter description length would require an inflation.

It is particular interesting to see this crucial impact of early developed genetic structures or modules on evolution.

1.5.5   Third experiment: Evolving artificial plants

Evolving artificial plants is an illustrative demonstration of the evolution of genetic representations to encode large-scale, structured, self-similar phenotypes.

The third experiment concerns the evolution of artificial plants in the encoding of Prusinkiewicz & Hanan (1989) that we already discussed in section 1.4.3 (page 76). The sequences we are evolving are arbitrary strings of the alphabet $\{A,..,P\}$ which are mapped on the L-system symbols according to $\{A,..,I\} \mapsto \{F,+,-,\&,\hat{},\backslash,/,[,]\}$ and $\{J,..,P\} \mapsto \{.\}$. Given such a sequence we evaluate its fitness by first drawing the corresponding 3D plant in a virtual environment (the OpenGL 3D graphics environment). We chop off everything of the plant that is outside a bounding cube of size $b \times b \times b$. Then we grab a bird's view perspective of this plant and measure the area of green leaves as observed from this perspective. The measurement is also height dependent: the higher a leave (measured by OpenGL's depth buffer in logarithmic scale where 0 corresponds to the cube's floor and 1 to the cube's ceiling), the more it contributes to the green area integral

$$L = \int_{x \in \text{bird's view area}} \Big[\text{color of } x = \text{green}\Big] \cdot \Big[\text{height at } x\Big] \frac{d\text{Area}}{b^2} \quad \in [0,1] \,.$$

$$(1.15)$$

Figure 1.8: Example for the weight functional. The dashed areas mark subtrees of the plant. The numbers indicate the weight each plant elements and subtrees. The plant's total weight is 13.

This integral is the positive component of a plant's fitness. The negative component is related to the number and "weight" of branch elements: To each element i we associate a weight w_i which is defined recursively. The weight of a leave is 1; the total weight of a subtree is the sum of weights of all the elements of that subtree; and the weight of a branch element i is 1 plus the total weight of the subtree that is attached to this branch element. E.g., a branch that has a single leave attached has weight $1 + 1 = 2$, a branch that has two branches each with a single leave attached has weight $1 + (2 + 1) + (2 + 1) = 7$, etc, see figure 1.8. The idea is that w_i roughly reflects how "thick" i has to be in order to carry the attached weight. The total weight of the whole tree,

$$W = \sum_i w_i \,,$$

gives the negative component of a plant's fitness. In our experiments we used

$$f = L \; - \; \varrho W$$

as the fitness function, where the penalty factor ϱ was chosen $\varrho \in \{0, 10^{-6}, 10^{-7}\}$ in the different experiments.

Details of the implementation.   Evolving such plant structures already gets close to the limits of today's computers, both, with respect to memory and computation time. Hence, we use some additional techniques to improve efficiency: First, we impose different limits on the memory resource that a genotype and phenotype is allowed to allocate, namely three: (1) The number of symbols in a phenotype was limited to be lower or equal than a number M_{max}. This holds in particular during ontogenesis: If the application of an operator results in a phenotype with too many symbols, then the operator is simply skipped. (2) The number of operators in a genotype is limited to be $\leq R_{\max}$. If a mutation operator would lead to more chromosomes, this mutation operator is simply skipped (no other mutation is made in place). (3) There is a soft limit on the number of symbols in a single chromosome: A duplication mutation is skipped if the chromosome already has length $\geq U_{\max}$. The limit though does not effect 2nd-type mutations; an inflative mutation $\pi \cdot p$ may very well lead to chromosomes of length greater than $U_{\max}$.

Second, we adopt an elaborated technique of self-adaptation of the mutation frequencies. We used the scheme similar to the self-adaptation of strategy parameters proposed by Bäck (1996). Every genome i additionally encodes two real numbers α_i and β_i. Before any other mutations are made, they are mutated by

$$\alpha_i \leftarrow \alpha_i \left(\mathcal{S}\, \mathcal{N}(0, \tau) + \tau' \right), \tag{1.16}$$
$$\beta_i \leftarrow \beta_i \left(\mathcal{S}\, \mathcal{N}(0, \tau) + \tau' \right),$$

where $\mathcal{S}\, \mathcal{N}(0, \tau)$ is a random sample (independently drawn for α_i and β_i) from the Gaussian distribution $\mathcal{N}(0, \tau)$ with zero mean and standard deviation τ. The parameter τ' allows to induce a pressure towards increasing mutation rates. After this mutation, α_i and β_i determine the mutation frequencies of 1st- and 2nd-type mutations respectively.

The A-trial: Demonstrating problems and features. Let us discuss four trials made with different parameters. Table 1.8 summarizes the parameter settings. The A-trial was one of our first experiments with the system. It assumes a recursion time $T = 5$, which actually leads to problems and therefore is replaced by $T = 1$ in later experiments. I chose to discuss the A-trial not because it exhibits a nice and efficient evolutionary dynamics but rather demonstrates typical problems and features that one needs to know to understand the system.

See figure 1.9. For the A-trial, the curves show some sudden change at generation $\sim$450 where the fitness, the number of phenotypic elements, and the total genome length explodes. The most significant curve in the graph after this explosion is the decaying genome size. We will find this phenomenon of explosion followed by decay of genome length also in other experiments and it typically indicates that the genomes in this period are too large and mutationally unstable. Hence,

	A-trial	B-trial	C-trial	D-trial	
δ, l	30/.5	↩	20/.3	↩	L-system angle δ, width l of the leave geometry
b	10	↩	20	↩	size of the bounding cube
T	5	↩	1	↩	stopping time of development
ϱ	0	↩	10^{-7}	10^{-6}	factor of the weight term W in a plant's fitness
α	.01	↩	↩	↩	(initial) frequency of first type mutations ($\tilde{a}$, aa, $\emptyset$)
β	.1	.01	.01	.3	(initial) frequency of second type mutations
τ (τ')	1. (0)	.2 (0)	.5 (.1)	0 (0)	rate of self-adaptation of α and β
	rank	μ, λ	↩	↩	type of selection
	yes	no	↩	↩	crossover turned on?
	$\langle A \rangle$	$\langle AFJ \rangle$	$\langle AAAFFFJ \rangle$	↩	initialization of $\Psi^{(0)}$ of all genotypes in the first population (they have no operators)
$\mathcal{A}$	{A,..,P}	↩	↩	↩	symbol alphabet
λ	100	↩	↩	↩	(offspring) population size
μ	30	↩	↩	↩	(selected) parent population size
$M_{\max}$	1000	10 000	100 000	1 000 000	maximal number of symbols in a phenotype
$R_{\max}$	100	↩	↩	↩	maximal number of operators allowed in one genotype
$U_{\max}$	40	↩	↩	↩	symbol duplication mutations are allowed only if a sequence has less than $U_{\max}$ symbols

Table 1.8: Parameter settings of the four trials. The "↩" means "same as for the previous trial" and shows that only few parameters are changed from trial to trial.

the innovation extincts and genomes are decaying until, in generation $\sim$1700, another innovation occurred and leads to a stable situation at generation $\sim$2600.

In table 1.9, the illustrations of the best individual in selected generations explain in more detail what happened. Until generation 420 the population consists of very basal phenotypic structures of no more than about 44 elements. Then, within the next few generations, a restructuring of the genome leads to much more complex phenotypes and the number of elements increases rapidly up to $\sim$600 elements in generation 465. However, this restructuring was such that all operators disappeared and the phenotypes were solely encoded by a huge $\Psi^{(0)}$; the genome size increases severely (from 37 in generation 420 to 2742 in generation 465) which leads to very high mutability. As a consequence these phenotypes are unstable, phenotypic variability becomes almost chaotic. The former innovation dies out, the genome size settles back to 73 in generation 1650 where the phenotype has only 25 elements. In generation 1680 a new innovation occurs. Again the recursive dependency of the new C-operator is exploited to encode a large number of phenotypic elements (326 in generation 1680). But this time the genome size remains reasonably limited (82) to ensure the stability of the encoding. This kind of genome dominates the next 1000 generations until, in generation 2560, a slight reorganization of that same genome leads to an even shorter genome that encodes about the same number of phenotypic elements in a very structured way—see the hexagonal structure. This genetic organization overtakes the generation at about generation 2560 and is continuously further optimized in the following. In the end, a second recursive G-operator was added and a very short genome of size 62 encodes a nicely hexagonally structured, dense phenotype of 322 elements that efficiently covers the area with green leaves.

What this trial demonstrates is the important interplay between genome size and phenotypic innovation. Only a properly organized genome can stably encode a large phenotype. Our choice of a recursion time $T = 5$ lead to the problem of very sudden phenotypic "explosions."

The B-trial: Fixing problems of the A-trial. The B-trial adopts most parameters of the A-trial except for the mutation rate parameters, axion initialization, and $M_{\max}$. The mutation rate parameters were chosen to make the evolution more stable, in particular the self-adaptation of mutation frequencies was damped ($\tau = .2$). The maximal phenotype size was increased to $M_{\max}=10\,000$ and all $\Psi^{(0)}$ were initialized by a small reasonable phenotype to speed up the early phase of evolution. We only briefly need to discuss the results. The second graph in figure 1.9 exhibits a clear structure. The gnome sizes are one order of magnitude lower than for the A-trial while the number of phenotypic elements encoded is an order of magnitude higher. At generation 200, the crucial innovation occurs and rapidly increases fitness almost to the optimal value 1.

The illustrations in table 1.10 confirm that evolution occurs rather orderly. Up to generation 180, phenotypes are improved in tiny steps. In generation 195 operator usage is exploited. The genome, consisting of the egg cell and one operator that is always applied 5×, is continuously optimized until it encodes surface-like phenotypic structures in generations 205 to 220. The concept is further developed until in generation 380 the plant's leaves cover almost all the area.

The C-trial. The major difference of the C- and D- trials to the A- and B-trials is that we set the recursion time T to 1 and we introduce a non-vanishing weight penalty factor $\varrho = 10^{-7}$. The consequence of $T = 1$ is that it is much harder for a genome to encode large phenotypic structures. Instead of a single operator that is applied 5×, the genome has to develop more operators, each of which is only applied once. As a consequence, the phenotypic structures become more interesting. The weight punishing factor ϱ enforces structures that are regularly branched instead of long curling arms. Further, we increase the limit $M_{\max}$ on the maximal phenotype size and initialize $\Psi^{(0)}$ non-trivially with small phenotypes to speed up early evolution.

In the third graph of figure 1.9 we also see some sudden change at generation $\sim$4000 where the fitness, the number of phenotypic elements, the number of operators in the genomes, and the total genome length explodes. Between generation $\sim$4000 and $\sim$5400, the most significant curve in the graph is the repeatedly decaying genome size. Indeed we will find that the genomes in this period are too large and mutationally unstable. The innovations extinct and genome size decays until at generation $\sim$5400 a comparably large number of phenotypic elements can be encoded by much smaller genomes that comprise more operators.

Referring to table 1.11 we find that for a long time not much happens until, in generation 4000, a couple of leaves turn up at once at certain places of the phenotype. This is exactly what we defined a correlated phenotypic adaptation and was enabled by encoding all the segments that now carry a leave within one operator. The concept is rapidly exploited until, in generation 4025, every phenotypic segment has a leave attached and is encoded by the single operator, namely the A-operator. The resulting "long-arm-building-block" triggers a revolution in phenotypic variability and leads to the large structures as illustrated for generations 4400 (3467 elements) and 4500 (7698 elements). However, as we had it in the A-trial, these structures are not encoded efficiently, the genome size is too large (512 and 720, respectively) and phenotypic variability becomes chaotic. The raise is followed by a fall of this species until, in generation 5100, evolution finds a much better structured genome to encode large phenotypes. The J-operator becomes dominant and allows to encode 1479 phenotypic elements with a genome size of 217. This beautiful concept is further improved and evolves until, in generation 8300, a genome of size 141 with 2 operators encodes a regularly structured phenotype of 3652 elements.

The D-trial. For the D-trial we turned off the self-adaptation mechanism for the mutation frequencies (based on the experience with the previous trials we can now estimate a good choice of $\alpha = .01$ and $\beta = .3$ for the mutation frequencies and fix it) and increase the limit M_{max} to maximally $1\,000\,000$ elements per phenotype. The severe change in the resulting structures is also due to the increase of the weight penalty factor ϱ to 10^{-6}—the final structure of the C-trial has a weight of about $.3 \cdot 10^{-6}$ which would now lead to a crucial penalty. The weight punishing factor ϱ enforces structures that are regularly branched instead of long curling arms.

Table 1.12 presents the results of the D-trial. Comparing the illustrations for generation 950, 1000, and 1010 we see that evolution very quickly developed a fan-like structure that is attached at various places of the phenotype. The fans arise from an interplay of two operators: The N-operator encodes the fan-like structures while the F-operator encodes the spokes of these fans. Adaptation of these fans is a beautiful example for correlated exploration. The N-operator encodes more and more spokes until the fan is complete in generation 1010, while the F-operator makes the spokes longer. Elongation proceeds and results in the "hairy", long-armed structures. Note that, in generation 1650, one N- and two B-operators are redundant. Until generation 1900, leaves are attached to each segment of the arms, similar to generation 4025 of the C-trial. At that time, the plant's weight is already $105\,099$ and probably prohibits to make the arms even longer (since weight would increase exponentially). Instead a new concept develops: At the tip of each arm two leaves are now attached instead of one and this quickly evolves until there are three leaves, in generation 1910, and eventually a complete fan of six leaves attached at the tip of each arm. In generation 2100, a comparably short genome with 10 used operators encodes a very dense phenotype structure of 9483 elements.

Conclusions. Let us briefly discuss whether similar result could have been produced with a more conventional GA that uses, instead of our non-trivial genotype-phenotype mapping, a direct encoding of sequences in $\{$F,+,-,&,^,\, /,[,]$\}$ that describe the plants. For example, setting $\beta = 0$ in our model corresponds to such a GA since no operators will be created and the evolution takes places solely on the "egg cell" $\Psi^{(0)}$, which is equal to the final phenotype in the absence of operators. We do not need to present the results of such a trial—not much happens. The obvious reason is the unsolvable dilemma of long sequences in a direct encoding: On the one hand, mutability must be small such that long sequences can be represented stably below the error threshold of reproduction; on the other hand mutability should not vanish in order to

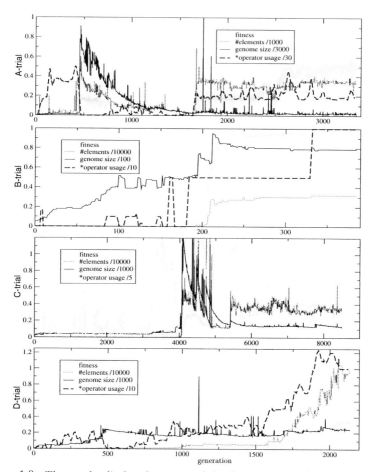

Figure 1.9: The graphs display the curves of the fitness, the number of phenotypic elements, the genome size, and the operators usage of the best individual in all generations of the four trials. Note that every quantity has been rescaled to fit in the range of the ordinate, e.g., the number of phenotype elements has been divided by 1000 as indicated by the notation "# elements/1000". The * for the operator usages indicates that the curve has been smoothed by calculating the running average over an interval of about a hundredth of the abscissae (30, 3, 80, and 20 in the respective graphs).

allow evolutionary progress. This dilemma becomes predominant when trying to evolve sequences of length $\sim 10^4$, as it is the case for the plants evolved in the D-trial. Also elaborate methods of self-adaptation of the mutation rate cannot circumvent this problem completely; the only way to solve the dilemma is to allow for an adaptation of genetic representations. The key novelty in our model that enabled the adaptation of genetic representations are the 2nd-type mutations we introduced.

In our example, two important features of the genetic representations coincide. First, this is the capability to find compact representations that allow to encode large phenotypes with small genotypes solving the error threshold dilemma. Second, this is the ability for complex adaptation, i.e., to induce highly structured search distributions that incorporate large-scale correlations between phenotypic traits. For example, the variability of one leave is, in certain representations, not independent of the variability of another leave. A GA with direct encoding would have to optimize each single phenotypic element by itself, step by step. The advantage of correlated exploration is that many phenotypic elements can be adapted simultaneously in dependence of each other.

Our experiments demonstrated the theory of σ-evolution which mainly states that the evolution of genetic representations is guided by a fundamental principle: They evolve such that the match between the evolutionary search distribution and the distribution of good solutions becomes better.

420: f=.031 e=44 w=421 o=15 g=37/3
$\Psi^{(0)} = \langle \text{HFAAOKAILEEJECELJJJJLJI} \rangle$
$\Pi = \langle \langle \text{O:OII} \rangle, \langle \text{L:IBJDAL} \rangle, \langle \text{F:FA} \rangle \rangle$

A simple phenotype where the recursive L-operator encodes the regular 5-leaves-fans and the recursive F-operator produces the long stem.

465: f=.21 w=604 e=9437 o=0 g=2742/0
$\Psi^{(0)} = \langle 2742 \rangle$
$\Pi = \langle \rangle$

Good fitness but all the phenotype is encoded in the huge axiom; mutability is too high.

1650: f=.032 e=25 w=223 o=0 g=73/0
$\Psi^{(0)} = \langle 73 \rangle$
$\Pi = \langle \rangle$

The previous innovation died out.

1680: f=.27 e=326 w=15361 o=5 g=82/1
$\Psi^{(0)} = \langle 74 \rangle$
$\Pi = \langle \langle \text{C:CCAIJDD} \rangle \rangle$

Similar fitness as in generation 465, but the regular phenotype is encoded by the recursive C-operator such that the genome size remains reasonable.

2560: f=.41 e=295 w=5195 o=9 g=56/4
$\Psi^{(0)} = \langle 38 \rangle$
$\Pi = \langle \langle \text{G:D} \rangle, \langle \text{E:AA} \rangle, \langle \text{L:BCG} \rangle,$
$\langle \text{C:KAICCJDD} \rangle \rangle$

The C-operator was further developed leading to an efficient hexagonal structure; other operators further decrease the genome size.

3280: f=.56 e=322 w=5887 o=6 g=62/4
$\Psi^{(0)} = \langle 42 \rangle$
$\Pi = \langle \langle \text{K:CJ} \rangle, \langle \text{E:ICC} \rangle, \langle \text{C:AICCJDD} \rangle,$
$\langle \text{G:GJDD} \rangle \rangle$

In addition to the C-operator, the new recursive G-operator makes the phenotype even more dense.

Table 1.9: The A-trial. The illustrations display the phenotypes at selected generations. The two squared pictures in the right lower corner of each illustration display exactly the bird's view perspective that is used to calculate the fitness: The lower colored picture displays the plant as seen from above and determines which area enters the green area integral in equation (1.15), and the upper gray-scale picture displays the height value of each element which enters the same equation (where white and black refer to height 0 and 1, respectively). Below each illustration you find some data corresponding to this phenotype: f=⟨fitness⟩ e=⟨number of elements⟩ w=⟨plant's total weight⟩ o=⟨number of used operators⟩ g=⟨genome size⟩/⟨number of chromosomes in the genome⟩. Below, the genetic system Π is displayed. In all cases except the first, $\Psi^{(0)}$ is too large to be displayed here and only the sequence length is given.

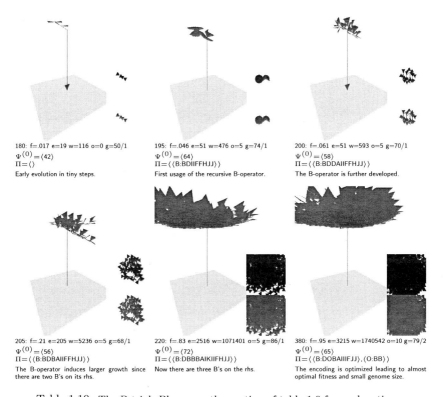

180: f=.017 e=19 w=116 o=0 g=50/1
$\Psi^{(0)} = \langle 42 \rangle$
$\Pi = \langle \rangle$
Early evolution in tiny steps.

195: f=.046 e=51 w=476 o=5 g=74/1
$\Psi^{(0)} = \langle 64 \rangle$
$\Pi = \langle \langle B:BDIIFFHJJ \rangle \rangle$
First usage of the recursive B-operator.

200: f=.061 e=51 w=593 o=5 g=70/1
$\Psi^{(0)} = \langle 58 \rangle$
$\Pi = \langle \langle B:BDDAIIFFHJJ \rangle \rangle$
The B-operator is further developed.

205: f=.21 e=205 w=5236 o=5 g=68/1
$\Psi^{(0)} = \langle 56 \rangle$
$\Pi = \langle \langle B:BDBAIIFFHJJ \rangle \rangle$
The B-operator induces larger growth since there are two B's on its rhs.

220: f=.83 e=2516 w=1071401 o=5 g=86/1
$\Psi^{(0)} = \langle 72 \rangle$
$\Pi = \langle \langle B:DBBBAIKIIFHJJ \rangle \rangle$
Now there are three B's on the rhs.

380: f=.95 e=3215 w=1740542 o=10 g=79/2
$\Psi^{(0)} = \langle 65 \rangle$
$\Pi = \langle \langle B:DOBAIIIFJJ \rangle, \langle O:BB \rangle \rangle$
The encoding is optimized leading to almost optimal fitness and small genome size.

Table 1.10: The B-trial. Please see the caption of table 1.9 for explanations.

3800: f=.0034 e=49 w=612 o=2 g=66/2
$\Psi^{(0)} = \langle 52 \rangle$
$\Pi = \langle \langle N:NNAA \rangle, \langle M:MMMPNNAA \rangle \rangle$

Over aeons not much happens.

4000: f=.0053 e=70 w=814 o=1 g=109/1
$\Psi^{(0)} = \langle 106 \rangle$
$\Pi = \langle \langle H:AA \rangle \rangle$

A first correlated phenotypic adaptation: the leaves along the stems pop up at once.

4025: f=.0087 e=156 w=1813 o=1 g=114/1
$\Psi^{(0)} = \langle 108 \rangle$
$\Pi = \langle \langle A:IMAJA \rangle \rangle$

Every stem segment is now encoded by the A-operator, which attached a leave to each segment.

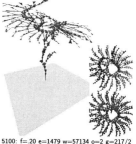

4400: f=.28 e=3467 w=92031 o=5 g=512/5
$\Pi = \langle \langle K:KA \rangle, \langle J:64 \rangle, \langle B:BF \rangle,$
$\langle G:GIFJCIJAIJA \rangle, \langle O:29 \rangle \rangle$

The concept of "long arms" is exploited, fitness explodes, but the genome becomes too large and non-stable.

4500: f=.26 e=7698 w=216119 o=3 g=720/3
$\Pi = \langle \langle K:O \rangle, \langle J:105 \rangle, \langle O:33 \rangle \rangle$

The large genome makes exploration chaotic; the species will extinguish.

5100: f=.20 e=1479 w=57134 o=2 g=217/2
$\Pi = \langle \langle K: \rangle, \langle J:77 \rangle \rangle$

A new concept with much shorter and stable genome takes over; the J-operator becomes dominant.

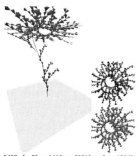

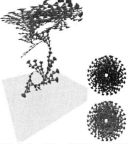

5400: f=.22 e=1410 w=59219 o=3 g=159/3
$\Pi = \langle \langle I:AJ \rangle, \langle C:JJCJ \rangle, \langle J:64 \rangle \rangle$

The new concept is further developed and exploited till the end ...

7500: f=.39 e=3215 w=350509 o=3 g=135/3
$\Pi = \langle \langle I:39 \rangle, \langle J:38 \rangle, \langle B:DDHDDJAIC \rangle \rangle$

8300: f=.31 e=3652 w=379288 o=2 g=141/2
$\Pi = \langle \langle I:39 \rangle, \langle J:60 \rangle \rangle$

Table 1.11: The C-trial. Please see the caption of table 1.9 for explanations. For some operators, the size of the rhs is given instead of the sequence.

950: f=.018 e=144 w=2681 o=3 g=163/3
Π=⟨⟨N:IP⟩,⟨P:JF⟩,⟨F:IIAJJF⟩⟩

Early evolution develops small phenotypes. Here, the interplay between the N- and the F-operator starts.

1000: f=.025 e=218 w=4337 o=2 g=171/2
Π=⟨⟨N:IJFFFFFFF⟩,⟨F:IIAJJF⟩⟩

The N-operator encodes fan-like structures attached at various places of the phenotype, the F-operator encodes the spokes of these fans.

1010: f=.032 e=506 w=10250 o=2 g=211/2
Π=⟨⟨N:NIJFFFFFFFFFF⟩,⟨F:IIAAJJF⟩⟩

Adaptation of these fans is a beautiful example for correlated exploration: the N-operator encodes more spokes; the F-operator makes them longer.

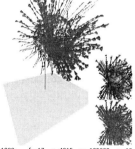

1650: f=.052 e=1434 w=31476 o=5 g=180/8
Π=⟨⟨B:IJNN⟩,⟨N:IAAJFIAAJF⟩,⟨N:NNBJ⟩, ⟨B:JJJ⟩,⟨B:NFNNKKKKB⟩,⟨F:IINFNNKK KCBJF⟩,⟨B:IJNNAJ⟩,⟨N:IJBA⟩⟩

The fan spokes become longer and longer. Note that one N- and two B-operators are redundant.

1750: f=.11 e=2867 w=65674 o=7 g=228/8
Π=⟨⟨B:NN⟩,⟨N:IAABFIAFJF⟩,⟨N:52⟩,⟨B:B BBJ⟩,⟨B:MMNNLDLCO⟩,⟨F:IIBJF⟩,⟨B:29⟩, ⟨N:NNNIJA⟩⟩

Leaves are attached to each segment of the long arms; similar to generation 4025 of the C-trial.

1900:   f=.17   e=4915   w=105099   o=10 g=230/12
Π=⟨⟨B:NN⟩,⟨N:IABFIAAJF⟩,⟨N:33⟩,⟨J:MO FJ⟩,⟨D:KNME⟩,⟨B:BBBBJ⟩,⟨B:FB⟩,⟨B:MN NLDDM⟩,⟨F:32⟩,⟨B:28⟩,⟨N:NLA⟩,⟨L:IJ⟩⟩

Two leaves are attached at the tip of each arm. The genome becomes to complex to be easily understood.

1910:   f=.20   e=4340   w=89996   o=12 g=226/14
Π=⟨⟨B:NN⟩,⟨N:IABFIAAJF⟩,⟨N:33⟩,⟨J:M MJ⟩,⟨J:NJ⟩,⟨D:KNME⟩,⟨B:BBBBJ⟩,⟨B:FB⟩, ⟨B:MNNLDDM⟩,⟨F:30⟩,⟨B:CCHEGEJFNJJ JMNLLCK⟩,⟨C:KCKKCLK⟩,⟨N:NLA⟩,⟨L:IJ⟩⟩

Now three leaves are attached at each tip.

1940:   f=.26   e=7366   w=149003   o=11 g=290/13
Π=⟨⟨B:NN⟩,⟨N:IAFBFIAAJF⟩,⟨N:85⟩,⟨J:M J⟩,⟨J:NJ⟩,⟨D:DONMME⟩,⟨B:FBFBFBFBFB J⟩,⟨B:MNNLDDM⟩,⟨F:34⟩,⟨B:CCJHMEJNJJ JMNLLCK⟩,⟨C:CCGKKCLKK⟩,⟨N:NNLA⟩, ⟨L:IJ⟩⟩

Now six leaves...

2100:   f=.33   e=9483   w=192235   o=10 g=261/15
Π=⟨⟨B:NN⟩,⟨N:IAABFIAAJF⟩,⟨N:57⟩, ⟨J:B⟩,⟨J:JJ⟩,⟨D:DOIE⟩,⟨B:FBFBFBFBBHJ⟩, ⟨B:ENNDD⟩,⟨F:36⟩,⟨B:25⟩,⟨B:BJEMLL⟩, ⟨C:CCGLB⟩,⟨B:CLK⟩,⟨N:NNLA⟩,⟨L:IJ⟩⟩

The plant becomes more and more dense and the genome size is optimized.

Table 1.12: The D-trial. Please see the caption of table 1.9 for explanations.

1.6   Conclusions

Throughout this chapter we developed a point of view on evolutionary adaptation that emphasizes the notion of the phenotypic exploration distribution, which is a general description of all of an individual's mutational properties. As a benefit we can formalize the relation and conceptual similarity between natural evolution, evolutionary algorithms, and the generic heuristic search scheme that we introduced as a basic information theoretic paradigm of what it means to "learn the structure of a problem and exploit it for future exploration." All three of them are based on a process of accumulating information—in a more or less explicit way. Concepts of self-adaptation or derandomized adaptation of probabilistic search strategies in evolutionary computation become comparable to the evolution of exploration strategies in natural evolution.

Based on this formalism we analyzed the implications of a non-trivial genotype-phenotype mapping for phenotypic search. The major result is the theorem on σ-evolution which basically proves that genetic representations, in particular also neutral traits in a neutral set, evolve such that the information given by selection is indeed accumulated: Genetic representations evolve such that the induced phenotypic exploration distributions (the "world models of evolution") adopt the structure of experienced selection distributions. This similarity has been termed the "required match between the functional relationships of the phenotypic characters [in the selection distribution] and their genetic representation [inducing the exploration distribution]." (Wagner & Altenberg 1996, section 6)

We also analyzed the meaning of crossover in terms of the structure of the exploration distribution it induces. The result is that crossover essentially *destroys* correlations between genes ("linkage correlations") by transforming the mutual information between the genes, that is present in the non-crossed population, into entropy of the crossed population. In that sense, crossover does the inverse of correlated exploration. Consequently, Holland's notion of building blocks significantly differs from our idea of functional phenotypic modules based on correlated exploration.

The discussion of natural genetic encodings (the DNA and its variability; transcription, translation and protein folding; gene regulation and the operon) showed that many aspects can be understood and systematized by discussing them in our formalism, i.e., always rethinking of what these mechanisms mean in terms of structuring the phenotypic exploration distribution. Especially the discussion of the operon, as a paradigm for the induction of correlated exploration, and the evolution of the tRNA molecules (which pinpoints the issue of "evolving GP-maps") proves how helpful it is to backtrack discussions to the central theme of structure in phenotypic exploration distributions and their σ-evolution.

The computational model was designed to demonstrate and illustrate what the theory on σ-evolution is about. We chose a genotype-phenotype mapping that incorporates basic mechanisms to induce correlated variability similar to the operon: Generically a genotype comprises an initial, premature phenotype ("the egg cell") and a set of transformation operators ("genes or operons") that recursively transform the phenotype during ontogenesis. The key point are the interactions between the operators that emerge from their recursive application, i.e., the operon like dependencies of their expression. This grammar-like GP-map is highly non-trivial in the strict sense we defined it; there exist genotypes that encode for the same phenotype but induce completely differently structured phenotypic variation distributions. Our model goes beyond similar models in the literature by introducing 2nd-type mutations that allow for neutral transitions between equivalent genotypes and thereby enable a neutral σ-evolution as demonstrated in the experiments. These 2nd-type mutations are neutral reformations of the genetic system, which result in a change of phenotypic exploration being not only a rescaling of the phenotypic neighborhood (as it is typically for strategy parameters) but a severe change of the variational topology on the phenotype space.

The model illustrates our claim that the underlying mechanism that originates the evolution of complex phenotypic adaptation is σ-evolution. Gene interactions emerge as a result of interdependent gene functionalities. Interesting is the strong impact of early developed genetic structures (modules) on future exploration and evolution. Some genomes also store neutral operators (genes that are not expressed during ontogenesis) which can be triggered non-neutral by a single mutation and thereby structurally influence phenotypic exploration. This literally demonstrates how the genetic representation of a phenotype, in particular also neutral parts of a genome, store information (measurable in terms of the exploration distribution) that has been accumulated earlier and can be exploited by further innovation.

The way genetic systems are organized is a mirror of what evolution has learned about the problem.

Chapter 2

Neural adaptation

2.1 Introduction

Decomposed adaptation and neural systems. The basis for the following work is to distinguish between the *functionality* of an adaptive neural system and its *parameters*. The functionality generally means the way of information processing of the network, which might be to map a given input on some response or to deliver a control signal. The parameters are the system's free variables that are directly subject to adaptation, namely synaptic (and bias) weights. This perfectly matches the notions of phenotype and genotype. As in the previous chapter, a central issue will be the analysis of the relation between parameters ("genotype") and functionality ("phenotype") and the influence of this relationship on the adaptation process.

To illustrate the approach reconsider the example about playing the piano and riding the bike that was mentioned in the introduction. At first sight, there is no way to say how these two functionalities are *represented* in the brain. But there is some indication if we also consider the adaptation process. Usually, when learning to play the piano one will not automatically unlearn (or even learn) to ride a bike and vice versa. Hence, to put it in the language of the previous chapter, there exists no *adaptational correlation* (cf. "no correlations in phenotypic exploration") between the two functionalities ("phenotypic traits"). Based on this adaptational decorrelation one may call these two functionalities *functional modules* (cf. "building blocks"), where the notion module is used in an abstract rather than architectural way: It does not presume that there really exist physically separable parts of the neural system that correspond to each functionality; instead it requires that these functionalities are parameterized in a modular way, as we will define it precisely later.

In turn, in this view it is not useful to define the notion of functional modularity for *non*-adaptive systems. Given a non-adaptive system's fixed functionality, the system's internal structure becomes irrelevant; the system is just a realization of a single point in the space of functionalities—and why should one associate a structuredness like modularity to a single point? The situation would be the same as for evolution which does not evolve: The genotypic representation of a phenotype becomes irrelevant.

The situation though changes if one considers an adaptive system the functionality of which is not fixed. Adaptation means a transformation, actually a translation of this point in functionality space. Over the time, the adaptive system realizes a *trajectory* through functionality space and it makes sense to analyze structural characteristics of this trajectory (like correlatedness in a stochastic path) as we will do. *The decomposability of a system is related to its behavior under adaptive transformation.*

Model Selection. The analysis of the space of functionalities and how neural systems parameterize this space is tightly related to the research on *model selection*. Following Kearns et al. (1995), the problem of model selection may be defined as follows: Given a finite set of data points, find a function (or conditional probability distribution, also called hypothesis) such that the expected generalization error is minimized. Here, generalization error means the error on a data subset that was *not* used to find or optimize the model. Typically, the search space F (the functionality space, e.g., the space of functions or conditional probability distributions) is assumed to be organized as a nested sequence of subspaces $F_1 \subseteq .. \subseteq F_d \subseteq .. \subseteq F$ of increasing complexity, see figure 2.1. For instance, the index d may denote the number of parameters or the Vapnik-Chervonenkis dimension (Vapnik 1995). Finding the function with minimal generalization error then amounts to finding the appropriate sub-search-space before applying ordinary optimization schemes. Many approaches to solve the model selection problem introduce a penalty term related to complexity which has to be minimized together with the training error in order to find the appropriate

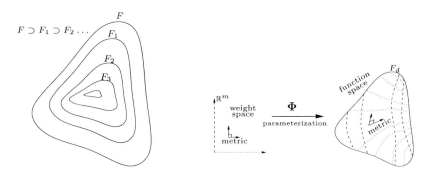

Figure 2.1: Two points of view on the model selection problem. On the left the functionality space F is organized as a nested sequence of subspaces $F_1 \subseteq .. \subseteq F_d \subseteq .. \subseteq F$ of increasing complexity, which illustrates the traditional approaches. On the right, the parameterization of a chosen functionality space F_d and the resulting induced functional metric is emphasized, what illustrates our approach.

model complexity. Penalty terms are, for example, the number of parameters of the model, the number of *effective* model parameters, the Vapnik-Chervonenkis dimension, or the description length (Akaike 1974; Amari 1993; Moody 1991; Rissanen 1978; Vapnik 1995). An alternative based on geometric arguments is presented by Schuurmans (1997).

As illustrated by figure 2.1, our approach has a different emphasis than the classical research on model selection. The choice of a specific model (e.g., a neural network) to represent a function has *two* implications: It defines the complexity and sub-space F_d of representable functions—as it is emphasized by classical model selection theories—but it also defines a *parameterization* of this space in the sense of introducing coordinates on that space, i.e., introducing a mapping $\Phi : W \to F_d$ from some parameter space W onto the respective sub-search-space. To avoid confusion, we use the term *model class* for the sub-search-space F_d, and *model parameterization* for the parameterization Φ of this sub-search-space. For example, an artificial neural network with m free parameters, fixed topology, and fixed activation functions defines a model class (the subspace of functions it can realize, which, if the topology is appropriate, includes an approximation of any function, Hornik, Stinchcombe, & White 1989)—but it also defines the model parameterization, i.e., the way functionalities are represented by the neural system.

As was mentioned in the introduction, a reason for the strong interest in model classes instead of model parameterizations might be the ultimate impact of the universal approximation theories as well as gradient learning. Given these fundamentals of classical Artificial Neural Network research, it seems that there is hardly motivation to rethink the way neural systems parameterize functions. Nevertheless, a non-structured, arbitrary parameterization of functionalities leads to problems:

Generalization, cross-talk, catastrophic forgetting, and the credit assignment problem.   The benefit of the analysis of the model parameterization is an understanding of the precise relation between variations of parameters and functional variations of the system. In turn, this is particularly helpful in understanding the adaptation dynamics of the system. In principal, adaptation occurs by variation of some parameters. In many cases, the observation of a single datum (a single learning step during online learning or a single reinforcement signal during Reinforcement Learning (RL)) triggers this adaptation. The model parameterization now describes the functional effects of the parameter adaptations—in particular also the variations of functionalities that were not subject to this single datum. Hence, the model parameterization is the key to understand the system's way of "generalizing" experienced to non-experienced data, or on how the system forgets previously learned data. This has indirectly been the subject of much research:

- Jacobs, Jordan, & Barto (1990) introduced the notion of spatial and temporal *cross-talk* which describes the fact that, in conventional neural networks, the response of two different neurons on the same datum or the response of the same neuron on two different data are typically not independent because the neurons share system parameters on which they depend. They argue that such crosstalk may be undesirable and is avoided by explicitly separating neurons in disjoint experts. As we will see below, selecting a multi-expert model is a very intuitive way to explicitly declare an independence of functional components and realize decomposed adaptation.

- *Catastrophic forgetting* describes that a conventional neural network, when trained for some time exclusively on new data, will typically forget the previously learned data (see French 1999 for a review). One may think of this as a special form of temporal cross-talk.

- Early approaches in Reinforcement Learning (before Q-learning became popular) used neural systems to directly implement the mapping between stimulus and response, or *policy* (see, e.g., Sutton & Barto 1998). At every time only a scalar feedback on the quality of the system output and not a precise feedback on the error of the output is given. Adaptation has to occur by some trial and error strategy. However, if the model parameterization is complex, it is very difficult to tell which parameter adaptations are responsible for which improvements or deteriorations of the system's functionality. This problem is called the *credit assignment problem*.

The connectionist way to parameterize functionalities. In this view, why should we use neural systems to represent functionalities in the first place? Besides the typical argument that nature does it, are there other principle reasons? In fact, many alternative systems are continuously proposed, most importantly Support Vector Machines, Bayesian Networks, etc. In some domains they exhibit better performances than neural systems and in other domains they are worse. In my view, the odds in favor of neural systems is that they in principal allow to adapt *the way* they parameterize functionalities. This "meta-adaptation" could be realized by evolutionary means or by neural mechanisms themselves (like gating mechanisms). The reason we argue for the adaptability of the way of parameterization goes back to the discussion of "complex adaptation mechanisms on arbitrary representations—or simple adaptation mechanisms on suitable representations" given in the introduction. We argued that current research tends to proceed with more and more sophisticated adaptation mechanisms on arbitrary representations instead of developing sophisticated mechanisms to adapt the representation such that simple functional adaptation—probably as simple as the Hebb rule—becomes possible. Here, the general advantage of connectionist systems might be that, due to their structural generality, they allow to implement any kind of modular, hierarchical, functionally dependent or independent representations.

Overview. In the next section we propose a theory, based on a differential geometry point of view, on the (co-)adaptation behavior of conventional feedforward neural networks (FFNNs). For instance, the theory allows to analytically describe the rate of forgetting of these networks which we exemplify and verify in a basic example. We apply the method of analyzing the model param-

eterization on a whole class of standard FFNNs. We find that the variety of
FFNNs with arbitrary topology is actually not a great variety with respect to
certain characters of the model parameterization. In particular, FFNNs generically introduce strong correlations between functional variations and thereby are
predisposed to forget previously learned data. Using FFNNs as a function model
means a limitation—not with respect to representable functions but with respect
to learning characteristics. A comparison of conventional neural networks with
networks that incorporate competitive interactions between neurons reveals how
competition in principle allows to structure the representation of functionalities
in favor of decorrelated adaptation dynamics.

These results have important implications for practical model selection: Commonly evolutionary algorithms are used to optimize network architectures in
favor of the system's adaptability. We argue that the search space of architectures should be generalized to include also systems that incorporate competitive interactions. We propose such a new class of systems in section 2.3
that may incorporate competitive as well as so-called gating interactions between neurons anywhere in their architecture. This class generalizes the class of
conventional neural networks as well as the class of conventional Multi-Expert
Systems. Four straight-forward learning schemes for these systems can be applied on any architecture: gradient learning, EM-learning, Q-learning, and a
variant of Q-learning. A selection of tests proves the functionality of these systems and learning schemes. Systems of this class can realize arbitrary correlated
or decorrelated representations of system functionality.

Parts of this work were first published in (Toussaint 2002b; Toussaint 2002a).

2.2 Functional modularity in NNs: Coadaptation and decomposed adaptation in neural systems

2.2.1 Decomposed adaptation

The adaptation dynamics of a neural system can be described as a Markov process—the structure of the transition distribution constitutes the notions of correlated adaptation and functional modules of a neural system.

As we did it for evolutionary systems, we define the adaptation process of neural systems on the level of system parameters. Instead of genes, synaptic and bias weights represent these degrees of freedom and we assume there are a finite number of real-valued weights $w \in W = \mathbb{R}^m$ that parameterize the neural system. We distinguish these system parameters from the *functional traits* of the neural system just as we distinguished genotype from phenotype. Let us first assume the simplest case where a neural system represents and input-output mapping $f : X \to \mathbb{R}^n$; below we will generalize this to probabilistic neural systems. In this case, functional traits refer exactly to this input-output mapping—from an external point of view only this input-output mapping is observable and may determine the quality of the system (cf. definition 1.2.5, page 32, of a phenotype). Hence, in analogy to the phenotype space, let us define F as the space of all mappings $X \to \mathbb{R}^n$; a given neural system with fixed parameters w then corresponds to exactly one point $\Phi(w) \in F$ in this space. The mapping $\Phi : W \to F$ is in analogy to the genotype-phenotype mapping and will be within our focus in the following considerations.

As argued above, our approach to define modularity is to consider the system's behavior under adaptation. More precisely, we want to characterize the correlation of different functional traits of the system under adaptation. Let us assume that the adaptation process is a Markov process described by a stochastic transition operator $\mathcal{H} : W \to \Lambda^W$ that acts on the system parameters, i.e., maps a current parameter state $w^{(t)}$ on the probability distribution $\mathcal{H}w^{(t)}$ of adapted parameter states. Stochastic adaptation dynamics are then given by

$$w^{(t+1)} = \mathcal{S}^1 \, \mathcal{H} \, w^{(t)} \,,$$

where the sample operator $\mathcal{S}^1$ describes the stochastic process of sampling the new weight configuration $w^{(t+1)}$ from the probability distribution $\mathcal{H}w^{(t)}$ (see

the definition on page 25). The basic example for stochastic adaptation is online
learning, where at each time a training datum referring to one or a few functional
traits is drawn independently from a data distribution.

The function space F, which could also be written as $(\mathbb{R}^n)^X$, is $n|X|$-dimensional
and the system functionality $\Phi(w)$ can be described by $n|X|$ real-valued numbers
f^a, $a = 1..n|X|$, that we call *function components*. These function components
f^a may also be regarded as entries of a lookup-table representation of the function $\Phi(w)$ where the index a specifies the location in the lookup-table. Namely,
an index a refers to a specific input x (say, the row of the lookup-table) and
an output dimension i (the column) such that f^a is the i-th dimension of the
system's output for some input x.

The question is whether two different functional traits *adapt in correlation*. We
capture this in the following definition:

Definition 2.2.1 (Adaptation covariance and coadaptation). Given a
neural model $\Phi : W \to F$ and an adaptation process $\mathcal{H} : W \to \Lambda^W$, we define
the *adaptation covariance* between two functional components a and b as

$$
\begin{aligned}
C_{ab}(w) &= \mathrm{cov}_{\mathcal{H}w}(\Phi^a, \Phi^b) \\
&= \langle \Phi^a \, \Phi^b \rangle_{\mathcal{H}w} - \langle \Phi^a \rangle_{\mathcal{H}w} \langle \Phi^b \rangle_{\mathcal{H}w} \\
&= \sum_{w'} (\mathcal{H}w)(w') \left(\Phi(w')^a \, \Phi(w')^b \right) \\
&\quad - \left[\sum_{w'} (\mathcal{H}w)(w') \, \Phi(w')^a \right] \left[\sum_{w'} (\mathcal{H}w)(w') \, \Phi(w')^b \right],
\end{aligned}
$$

where $\langle \, \cdot \, \rangle$ means averaging over the specified distribution. The adaptation
covariance depends on the current parameter state $w \in W$ of the neural system.

The adaptation covariance is a very rich description of the adaptation process.
On the one hand, if the adaptation covariance between two functional traits is
non-vanishing, it describes how the neural system generalizes the adaptation of
one functional trait to another: For example, if during online learning, f^a is
tested and results in system adaptation, the covariance describes how another
component f^b is *co*-adapted in correlation to f^a, although it was not explicitly
subject to experience during the training. Whether this coadaptation is desirable
or not depends on the problem. In general, one would like to choose from a
variety of different coadaptation schemes, i.e., one would like to select a model
from a variety of models with different kinds of coadaptation.

On the other hand, if the adaptation covariance between two functional traits
f^a and f^b vanishes, the system will not generalize an adaptation of f^a to a
coadaptation of f^b. In that way, all the functional traits of the system may be
organized in groups; traits of the same group depend on each other and form a
functional module, traits of different groups adapt independently of each other:

Definition 2.2.2 (Adaptation decomposition and functional modules).
Let A be the index set of all functional traits. (For input-output functions
$X \to \mathbb{R}^n$, A has cardinality $n|X|$.) A partition of functional traits is given by
disjoint, non-empty subsets $A_1, .., A_g \subset A$ which unite to A, i.e. $\bigcup_{i=1}^g A_i = A$.
If there exists a partition such that the adaptation covariance becomes a block
matrix, i.e.,

$$\forall w \in W : \ a \in A_i, \ b \in A_j, \ i \neq j \implies C_{ab}(w) = 0 \,,$$

then we speak of *adaptation decomposition*. We call the collection of functional
traits f^a in one group $a \in A_i$ a *functional module*.

This can be transfered also to probabilistic models $X \to \Lambda^Y$ that map some input
to an output distribution over some space Y: The simplex Λ^Y of distributions is
a subspace of $\mathbb{R}^{|Y|-1}$, every distribution in Λ^Y can be parameterized by $|Y| - 1$
real numbers. Thus, a model $f \in F : X \to \Lambda^Y$ can likewise be described by
$(|Y| - 1)|X|$ real-valued functional components f^a.

A more information theoretic notion of correlated adaptation. The
covariance matrix C_{ab} can be interpreted in a more information theoretic style
when assuming that the *functional* adaptation distribution $\Xi \mathcal{H} w^{(t)} \in \Lambda^F$ is of
the exponential family, i.e.,

$$(\Xi \mathcal{H} w^{(t)})(f) = \frac{1}{C} \exp\left[- \sum_{ab} C_{ab} \, f^a f^b \right]$$

where the scaling factor C is uniquely given via normalization. The projection
$\Xi : \Lambda^W \to \Lambda^F$ of distributions over parameter space to distribution over the
functional space is defined in analogy to definition 1.2.8 (page 33). It can be
shown that the mutual information between two functional traits is then given
by

$$I^{ab} = \ln \frac{C_{aa} \, C_{bb}}{C_{aa} \, C_{bb} - C_{ab}^2} \, .$$

Hence, mutual information vanishes if and only if the adaptation covariance C_{ab} between these two traits vanishes.

2.2.2 The geometry of adaptation

The relation between a neural system's free parameters and its functional traits can be described geometrically. The induced metric on the functional space determines the correlatedness of adaptation.

The mapping Φ establishes the relation between the parameter space W and the functional space F. We introduced adaptation as a process on the parameter space that results in a trajectory on the parameter space. The mapping Φ tells us how the local properties of the trajectory, in particular its direction, translate to the adaptational direction in the functional space. This is in perfect analogy to our point of view that the genotype-phenotype mapping represents a lift of topology from a locality in genotype space to a corresponding locality in phenotype space.

In terms of differential geometry (see figure 2.2) the mapping Φ is (the inverse of) a coordinate map (or chart, or atlas) on the manifold of representable functions $\Phi(W) = \{\Phi(w) \mid w \in W\} \subseteq F$, i.e. the image of Φ. Such a mapping $\Phi : W \to F$ induces also the linear differential mapping $d\Phi(w) : T_w W \to T_{\Phi(w)} F$ from the local tangent $T_w W$ space of W to the corresponding tangent space of F. The differential map allows to lift or pull-back all objects (vectors, tensors, metrics) that live in one tangent space to the other. Above all, this allows to translate a vector $\Delta w \in T_w W$ that describes a variation of system parameters into a vector $\Delta\Phi = d\Phi(\Delta w) \in T_{\Phi(w)} F$ (sometimes also written $\Delta\Phi = \Delta w \rfloor d\Phi$) that describes the corresponding variation of system functionality. In coordinates this simply reads:

$$\Delta\Phi(w)^a = \sum_i \frac{\partial \Phi(w)^a}{\partial w^i} \Delta w^i \ .$$

In this language, we define a metric that will be a generic tool in understanding the relation between the variation of parameters and functional variations.

Definition 2.2.3 (Functional metric). Given a neural model $\Phi : W \to F$ and a metric $\hat{g}_{ij}$ on the parameter space, we define the *functional metric* on the

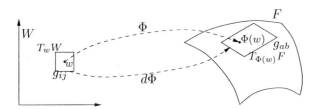

Figure 2.2: The differential geometry point of view. The parameterization Φ maps parameter states w on the system functionality $\Phi(w)$. Important for adaptation dynamics is the relation between parameter variations and the functional variations they induce. This relation is, in first order, given by the differential $d\Phi$ that maps from the tangent space $T_w W$ around w to the tangent space $T_{\Phi(w)} F$ around $\Phi(w)$. In particular it also allows to define the functional metric g_{ab} on the functional space which is induced by a presumed metric g_{ij} on parameter space.

(dual) tangent space $T^*_{\Phi(w)} F$ of F as

$$g^{ab}(w) := \sum_{ij} \hat{g}^{ij} \frac{\partial \Phi(w)^a}{\partial w^i} \frac{\partial \Phi(w)^b}{\partial w^i} \, , \tag{2.1}$$

where $\hat{g}^{ij}$ is the inverse of $\hat{g}_{ij}$. The inverse g_{ab} of g^{ab} is the functional metric on the ordinary tangent space $T_{\Phi(w)} F$.

2.2.2.1   Ordinary gradient online learning

We describe online learning within our formalism as follows: At every time, a training datum t^a for one functional trait a is drawn independently from a data distribution. The error of the system's functional component $\Phi(w)^a$ is measured with a functional $E(t^a, \Phi(w)^a)$. In the case of ordinary gradient learning with adaptation rate α, the adaptation of the system's parameters is then realized by

$$\Delta w^i = -\alpha \sum_a \frac{\partial \Phi(w)^a}{\partial w^i} \, e(w)_a \, , \quad e(w)_a := \frac{\partial E(t^a, \Phi(w)^a)}{\partial \Phi(w)^a} \, . \tag{2.2}$$

The components $e(w)_a$ are the dual components of the vector of steepest descent $e(w) = \sum_a e(w)^a \boldsymbol{\xi}_a = \sum_{a,b} \hat{g}^{ab} e(w)_b \, \boldsymbol{\xi}_a$ on the functional space, where $\boldsymbol{\xi}_a$ are the local basis vectors of the coordinate frame in F. E.g., for the square error, which corresponds to the Euclidean metric, $e(w)_a = 2 \, (t^a - \Phi(w)^a)$ is the actual error made at the component a. Being familiar with differential geometry, one realizes here already that equation (2.2), although it dominated NN research for decades, is in some sense confused because the lhs are (contravariant) components of a vector whereas the rhs are (covariant) components of a dual vector. One could say that ordinary gradient online learning hides the presumption of a fixed Euclidean metric on the parameter space W. To account for this presumption at least notationally, we rewrite the equation as:

$$\Delta w^i = \alpha \sum_j \hat{g}^{ij} \sum_a \frac{\partial \Phi(w)^a}{\partial w^j} e(w)_a \ , \quad \text{where } \hat{g}^{ij} \equiv \hat{g}_{ij} \equiv \delta_{ij} \text{ is Euclidean .}$$

$$(2.3)$$

On the functional level, this leads to an adaptation of all functional traits according to

$$\Delta \Phi(w)^b = \sum_i \frac{d\Phi(w)^b}{dw^i} \Delta w^i$$

$$= \alpha \sum_a \sum_{ij} \hat{g}_{ij} \frac{d\Phi(w)^b}{dw^i} \frac{d\Phi(w)^a}{dw^j} e(w)_a \ .$$

Applying definition (2.1) of the functional metric, this reads

$$\Delta \Phi(w)^b = \alpha \sum_a g^{ab} e(w)_a \ . \tag{2.4}$$

There are two points to emphasize when interpreting this simple equation:

- The lhs $\Delta \Phi(w)^b$ are the components of the vector $\Delta \Phi = \sum \Delta \Phi(w)^a \, \xi_a$ that describes the *actual variation* of the system functionality in the tangent space $T_{\Phi(w)} F$. In the rhs, $e(w)_a$ are the dual components of the vector of *steepest descent* in the dual tangent space $T^*_{\Phi(w)} F$. Now, equation (2.4) relates $\Delta \Phi$ to e "as if g^{ab} was the true metric on F". But g^{ab} actually depends on the choice Φ of model since it is derived from the presumed $\hat{g}_{ij}$ via (2.1). Definitely g^{ab} does not equal to a metric that one would *naturally* assume on F, e.g., the mean square metric on function space or the Fischer metric on probability distributions. Hence, for ordinary gradient descent, the actual direction of system adaptation is model dependent and does not equal to the natural steepest descent direction on F (i.e., the

steepest descent with respect to a naturally chosen metric on F.) E.g.,
it is not invariant under transformations of the parameter space. All this
stems from the ad hoc assumption of the Euclidean metric $\hat{g}^{ij} = \hat{g}_{ij} = \delta_{ij}$
on W in equation (2.2) respectively (2.3).

- Concerning the topic of coadaptation, we find that the functional metric
  g^{ab} describes the variation of a functional component $\Phi(w)^b$ when t^a is
  trained. In other words, *the functional metric g^{ab} is a first order description of how the neural system generalizes the experience of a target value
  t^a in order to adapt also functional components Φ^b.* This also allows to
  derive an exact expression for the adaptation covariance:

**Theorem 2.2.1 (Coadaptation and functional modules in ordinary gra-
dient online learning).** *In the case of ordinary gradient online learning, the
adaptation covariance is given by*

$$C_{bc} = 4\alpha^2 \sum_a p(a)\, g^{ba}\, g^{ca}\, \big(e(w)_a\big)^2 - \langle \Delta f^b \rangle \langle \Delta f^c \rangle \ .$$

*If $\langle f^a \rangle = 0$ (i.e., after the early phase of mean adaptation) then, for two traits
a and b, the adaptation covariance C_{ab} vanishes for arbitrary $e(w)$ if and only
there exists and ordering of indices such that functional metric g^{ab} can be written
as a block matrix and the indices a and b correspond to different blocks. These
blocks of the functional metric then define the functional modules of the system.*

Proof.

$$\begin{aligned}
C_{bc} &= \langle \Delta f^b \, \Delta f^c \rangle - \langle \Delta f^b \rangle \langle \Delta f^c \rangle \\
&= \sum_a p(a)\, 2\alpha\, g^{ab}\, 2\alpha\, g^{ac}\, \big(e(w)_a\big)^2 - \langle \Delta f^b \rangle \langle \Delta f^c \rangle \\
&= 4\alpha^2 \sum_a p(a)\, g^{ba}\, g^{ca}\, \big(e(w)_a\big)^2 - \langle \Delta f^b \rangle \langle \Delta f^c \rangle \ .
\end{aligned}$$

Concerning the first term, the product $g^{ba}\, g^{ca}$ vanishes for all a if and only if
there exists and ordering of indices such that the functional metric is a block
matrix and b and c refer to different blocks:

$$g = \left(\begin{array}{cc} A \in \mathbb{R}^{\mu \times \mu} & 0 \\ 0 & B \in \mathbb{R}^{\nu \times \nu} \end{array} \right) \ , \quad b \leq \mu \ , \quad c > \mu \ ,$$

where A and B are arbitrary symmetric matrices and $\mu + \nu = n|X|$. Thus, adaptation is decomposed into two subsets of functional components exactly if the functional metric is a block matrix and the functional component subsets correspond to these blocks (cf. equation (2) in the introduction).   □

2.2.2.2 Natural gradient online learning

We pointed out that in equation (2.2), the rhs $\alpha \sum_a \frac{d\Phi(w)^a}{dw^i} e(w)_a$ is actually covariant (with lower index i) whereas the lhs Δw^i is contravariant (referring to components of the parameter vector), and thus the equation is not invariant under transformations which means that, if one transforms the parameter space (e.g., a transformation from Cartesian to polar coordinates in W) then the real direction of the parameter adaptation vector $\Delta \boldsymbol{w}$ in the tangent space $T_{\Phi(w)}W$ changes ($\Delta \boldsymbol{w} = \sum_i \Delta w^i \boldsymbol{e}_i$, where $\boldsymbol{e}_i$ are the basis vectors of the coordinate frame) and thus also the real direction of the functional adaptation vector $\Delta \boldsymbol{\Phi}(w)$ changes. Amari (1998) was the first to notice this circumstance. He proposed to presume a meaningful—*natural*—metric $\hat{g}_{ab}$ on F instead of presuming the Euclidean metric $\hat{g}_{ij}$ on W and use a transformation invariant adaptation rule based on this metric.

There are two typical choices for the natural metric $\hat{g}_{ab}$ on F: If F is a space of functions, the mean square metric is the natural one; and if F is a space of conditional probability distributions, then the Fisher metric is the natural choice.

Given the natural metric $\hat{g}_{ab}$ on F, *natural gradient descent* is given by

$$\Delta w^i = \alpha \sum_j g^{ij} \sum_a \frac{d\Phi(w)^a}{dw^j} e(w)_a \ , \tag{2.5}$$

$$\text{where } g_{ij}(w) := \sum_{a,b} \hat{g}_{ab}(w) \frac{\partial \Phi(w)^a}{\partial w^i} \frac{\partial \Phi(w)^b}{\partial w^j} \ , \quad g^{ij} = \left[g_{ij} \right]^{-1} .$$

We see that natural gradient descent reads almost the same as ordinary gradient descent as given in (2.3) except for the important fact that g_{ij} is not presumed ad hoc and fixed independently of the parameterization but rather derived from the natural metric $\hat{g}_{ab}$ on F. In fact, g_{ij} is roughly the image of g_{ab} under $(d\Phi)^{-1}$, see figure 2.2, which is called "pull-back" in differential geometry.

Consequently, adaptation on the functional level yields

$$
\begin{aligned}
\Delta\Phi(w)^b &= \sum_i \frac{d\Phi(w)^b}{dw^i}\,\Delta w^i \\
&= \alpha \sum_a \sum_{ij} g^{ij}\,\frac{d\Phi(w)^b}{dw^i}\,\frac{d\Phi(w)^a}{dw^j}\,e(w)_a \\
&= \alpha \sum_a \hat{g}^{ab}\,e(w)_a \ .
\end{aligned}
$$

We find that the functional adaptation vector $\Delta\Phi(w) = \sum_a \Delta\Phi(w)^a\,\xi_a$ points exactly in the direction of the vector of steepest descent $e(w) = \sum_a e(w)^a\,\xi_a = \sum_{a,b}\hat{g}^{ab}\,e(w)_b\,\xi_a$ w.r.t. the chosen natural metric $\hat{g}_{ab}$ on F.

What does that mean for the discussion of coadaptation? Well, all the results that we have derived for ordinary gradient descent are valid here if we replace g_{ab} by the natural functional metric $\hat{g}_{ab}$. This holds in particular for theorem 2.2.1 and hence, the choice of the natural metric directly determines the system's coadaptation behavior. For instance, if the mean square metric is chosen as the natural metric on a function space F, it follows directly that coadaptation vanishes for $\langle f^a \rangle = 0$ (i.e., after the early phase of mean adaptation). The same is true if the system realizes a conditional probability $X \to \Lambda^{\{0,1\}}$ (i.e., f^a directly equals the probability $p(1|x)$ for some input $x \in X$) and the cross-entropy is chosen as the natural metric on this space F of conditional probability distributions. In both cases the metric on F is diagonal and coadaptation vanishes.

A note on the natural gradient. At first sight, these results seems very encouraging for the natural gradient. It shows that, using the natural gradient, one has no first order coadaptational effects, no catastrophic forgetting, no crosstalk. However, this might not always be advantageous. For some problems, coadaptation might be desirable and speed up learning—e.g., when the problem allows to generalize knowledge from one datum to another. Of course, one may argue that it is better to have, to first-order, no coadaptation instead of uncontrolled arbitrary coadaptation as with the ordinary gradient. But how could one achieve an adaptation of coadaptation, as it is done in the case of ordinary gradient learning by means of evolutionary architecture adaptation, in the case of natural gradient learning?

There exists a second, probably more obvious objection against the natural gra-

dient: its computational complexity. Natural gradient learning requires to com-
pute the inverse parameter metric g^{ij} and then multiply it to the covariant
derivative vector $\sum_a \frac{d\Phi(w)^a}{dw^j} e(w)_a$, see equation (2.5). If g^{ij} cannot be given analytically, there exists a method to approximate this metric "on the fly" (Amari,
Park, & Fukumizu 2000). However, all these calculations and also memory re-
sources scale with the size of the metric g^{ij}, i.e., with the square of number of
parameters. Even for small neural systems with about a hundred synapses, this
computational disadvantage might likely exceed the advantage in learning speed.
Large-scale neural systems with thousands of neurons can hardly be handled in
this way.

These are the reasons why, despite its theoretical elegance, we do not consider
any further natural gradient learning in the following.

2.2.3 An example: Comparing conventional NN with multi-expert models

*An example demonstrates how the theory precisely describes the difference
in adaptation dynamics between a conventional NN and a multi-expert
w.r.t. their coadaptation behavior.*

The task. In this section we illustrate the implications of the theory with a
basic example. The task we choose to test learning behavior is very simple.
A mean square error (MSE) regression of only two 3-bit patterns has to be
learned by mapping the first pattern on $+1$ and the second on -1. However,
we impose that these patterns have to be learned *online* where they alternate
only after they have been exposed for 100 times in succession. This task focuses
on the analysis of the coadaptation of the responses on these two patterns and
the 100-fold repetition of the same datum makes it non-trivial for conventional
artificial neural networks to learn it. The task is not unrealistic; similar effects
of learning and unlearning occur in online learning when a specific response is
unlearned during the course of training other responses for several time steps.
In real world simulations it is plausible that stimuli remain unchanged for many
time steps. The two patterns were chosen as 110 and 010. Learning is realized
by a slow gradient descent with adaptation rate $2 \cdot 10^{-3}$ and momentum 0.5.
The metric components are calculated from the gradients.

• The feed-forward neural network we investigate here is 3-4-1-layered; layers are completely connected; the output neuron is linear, the hidden ones implement the sigmoid $\frac{1}{1+\exp(-10\,x)}$; only the hidden neurons have bias terms.

• The softmax model is the same as the standard model with the exception that the four neurons in the hidden layer *compete* for activation: their output activations y_i are given by

$$y_i = \frac{e^{30\,x_i}}{X} \, , \quad x_i = \sum_{j \in \text{ input}} w_{ij} y_j + w_i \, ,$$

$$X = \sum_{i \in \text{ hidden}} e^{30\,x_i} \, . \tag{2.6}$$

Here, w_{ij} and w_i denote weight and bias parameters. The exponent factor 30 may be interpreted as a rather low temperature, i.e., high competition. The calculation of the gradient is a little more involved than ordinary back-propagation but straightforward and of same computational cost. (It is a special case of the general gradient we calculate in section 2.3.1.)

Table 2.1: The two models we investigate: A standard feed-forward neural network and a similar network involving a softmax competition.

The architectures. We test two neural systems on this task. On the one hand, this is a standard 3-4-1 feed-forward neural network and on the other, this is the same architecture with a softmax competition in the hidden layer (the details can be found in table 2.1). The parameters of both systems are initialized randomly by the normal distribution $\mathcal{N}(0, 0.1)$.

Results. Please see figures 2.3 and 2.4 for the results. The standard neural model exhibits strong forgetting of the untrained pattern during the training of the other. In contrast, for the softmax model the error of the untrained pattern hardly increases. The rate of forgetting, given by the slope of the error curve of the untrained pattern, is perfectly described by theory as given by equation (2.4); the graphs in the middle display these curves and show the perfect match of theory and experiment. The bottom graphs display the functional metric components and generally show that the cross-component g^{01}, which is

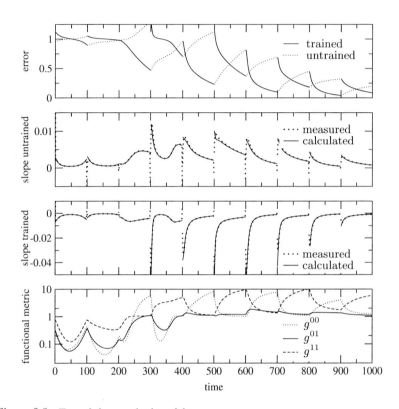

Figure 2.3: *Test of the standard model.*
For all four graphs the abscissa denotes the time step.
Top: The learning curves (errors) for both patterns are displayed. Only one of the patterns is trained—alternating every 100 time steps. The error of the untrained patterns increases.
Second: The slope (change of error per time step) of the untrained learning curve is displayed. The dotted line refers to the measured slope of the upper curve, the normal line is calculated according to equation (2.4).
Third: The slope (measured and calculated) of the trained learning curve.
Bottom: The three components of the functional metric g^{00}, g^{01}, g^{11} are displayed in logarithmic scale. In particular the cross-component g^{01} is clearly non-vanishing.

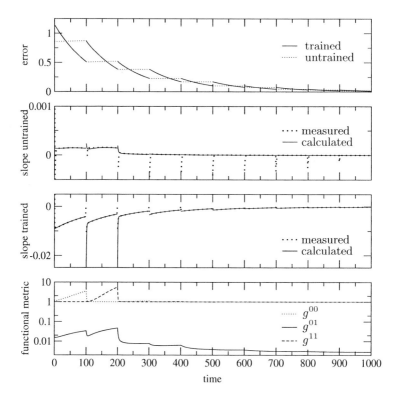

Figure 2.4: *Test of the softmax model.*
Top: The learning curves (errors) for both patterns are displayed. The untrained
patterns is scarcely forgotten.
Second: The slope (measured and calculated) of the untrained learning curve nearly
vanishes.
Third: The slope (measured and calculated) of the trained learning curve.
Bottom: The three components of the functional metric g^{00}, g^{01}, g^{11} (in logarithmic
scale). The cross-component g^{01} is small, it decreases significantly at time step 200.

responsible for coadaptation and forgetting, is quite large for the standard model compared to the softmax model. Further, the softmax model seems to learn the adaptation decomposition, as defined in section 2.2.1, after the 200th time step.

All these results demonstrate the significance of the developed theory on coadaptation. Not surprisingly, the standard model is not well-suited to solve the simple task given. Remarkably is that the components g^{00} and g^{11} become significantly greater than 1 during the training phase of the respective functional component. By equation (2.4), this means that the "effective" adaptation rate is larger than $2 \cdot 10^{-3}$.

2.2.4 Generic properties of the class of conventional NN and multi-expert models

The statistics of coadaptational features in the space of all neural networks of a given architecture and in the space of all multi-expert systems of a given architecture exhibits a generic difference.

The theory clarifies and emphasizes the meaning of the functional metric as a description of a system's coadaptation behavior; and the previous example gave some intuition about this metric. We can now generically analyze both types of neural systems, with and without competitive interactions, solely based on a discussion of the systems' functional metric.

We investigate the distribution of the functional metric components over the whole respective model classes. The two classes correspond to the two systems we investigated above, where the system architecture is fixed, but the parameters are free and span the class space. For both classes we want to calculate the probability that the functional metric components acquire a certain value. In other words, if we randomly choose a system from one or another class, what coadaptational feature may we expect from the system? To calculate these probabilities we need to assume a probability distribution over parameters for both classes; we will choose the normal distribution $\mathcal{N}(0, 0.1)$.

Figure 2.5 displays the distributions of functional metric components for both classes. Clearly, the standard model exhibits a Gaussian like distribution of the cross-component g^{01} with mean around 1.5; a vanishing cross-component g^{01} is not likely. On the other hand, the softmax model exhibits two strong peaks at $g^{01} = 0$ and $g^{01} = 1$, such that the probability for $g^{01} < 0.1$ is larger than 10%.

These distributions are generic properties of the two system classes and we may conclude:

- In the class of conventional feed-forward neural networks of the given architecture and with ordinary gradient descent it is very unlikely to have no coadaptation. *Coadaptation, cross-talk, and forgetting are generic and inevitable "features" of the class of feed-forward neural networks.*

- In the class of systems of the same architecture but with the competitive hidden layer, systems with non-vanishing and with identically vanishing

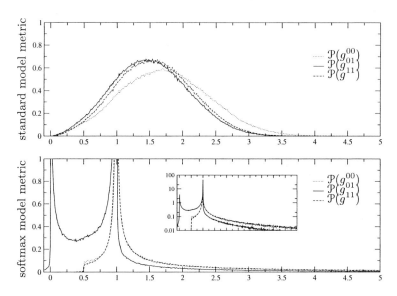

Figure 2.5: *Distribution of metric components.*
The distribution was calculated as a histogram of 1 million samples by using bins of
size $\frac{1}{100}$. All curves integrate to 1.
Top: The standard model. The probability of vanishing cross-component g^{01} is very
small.
Bottom: The softmax model. The inset graph is in logarithmic scale. The probability
of vanishing cross-component g^{01} is fairly high.

coadaptation are equally likely. The general point here is that *the class of
systems that incorporate competition has a greater variety w.r.t. coadaptational features.* Hence, if one aims at the adaptation of coadaptational
features, this class seems highly more suited as a search space than the
class without competition.

2.2.5   Summary

The cornerstones of this section are the three definitions made. First, the definition 2.2.1 (page 116) of coadaptation—in analogy to correlated evolutionary
exploration—allows a general description of adaptation dynamics with respect

to cross-talk, forgetting, or the way of generalization. The definition 2.2.2 (page 117) of functional modules transfers the notion of functional phenotypic modules from evolution to neural systems. And the definition 2.2.3 (page 118) of the functional metric, motivated by a differential geometric description of neural systems, allows to analyze coadaptation properties of neural networks. Based on this formalism, it was straightforward to derive the theorem 2.2.1 (page 121) on the coadaptation in neural networks.

The empirical investigations we made demonstrate the meaning of coadaptation and the functional metric as it is described theoretically. The last study offers a general characterization of conventional neural networks and systems with competitive interactions by analyzing the distribution of the functional metric over the whole model class, see figure 2.5. We concluded that one should generalize the search space of artificial neural systems to include also systems with competitive interactions in order to allow for an adaptation of coadaptation features. We propose such a generalization in the next section.

2.3   A space of more modular neural systems

The question of *how* to select a system with appropriate coadaptation features has not yet been addressed. As discussed in the introduction, classical approaches to model selection commonly introduce a penalty term in order to reduce the model's complexity. Following this tradition we could introduce a penalty term that reduces coadaptation. For instance, this could be the term $\sum_{ab}(g^{ab})^2 - \sum_a(g^{aa})^2$ which is a measure of the cross-components in the functional metric. Introducing this penalty term in the quality measure during (e.g., evolutionary) search for good systems would lead to a bias toward systems with reduced coadaptation.

Although this would perfectly match the classical approaches to model selection, this is not quite what we intended. Our goal is not to enforce a special kind of coadaptation but rather to allow for an adaptation of a system's coadaptation behavior to the specific problem. If certain coadaptational dependencies between functional components simplify the problem solution, then we should not enforce a bias towards vanishing coadaptation.

The evolution of neural networks, as it recently became an elaborated branch of research (Yao 1999), is a promising method for model selection, and specifically for the adaptation of coadaptational features. However, most of these existing approaches focus on the search space of conventional neural networks. The belief is that the variety of topologies offers a variety of functionally different models. The previous section challenges this belief, though, and motivates the generalization of the search space to also include systems with more structured functional metric. The presented softmax model involving competitive interactions between neurons is a step in this direction. In this section we will define a search space of neural systems that may comprise competitive and so-called gating interactions anywhere in the architecture. The resulting space of neural system generalizes conventional neural networks as well as multi-expert systems.

2.3.1   A neural model for multi-expert architectures

A generalization of conventional artificial neural networks allows for a functional equivalence to multi-expert systems.

When using multi-expert architectures for modeling behavior or data, the motivation is the separation of the stimulus or data space into disjoint regimes on which separate models (experts) are applied (Jacobs 1999; Jacobs, Jordan, & Barto 1990). The idea is that experts responsible for only a limited regime can be smaller and more efficient, and that knowledge from one regime should not be extrapolated to another regime, i.e., optimization on one regime should not interfere with optimization on another. In the previous section we have shown that this kind of adaptability cannot be realized by a single conventional neural network.

To realize a separation of the stimulus space one could rely on the conventional way of implementing multi-experts, i.e., allow neural networks for the implementation of expert modules and use external, often more abstract types of gating networks to organize the interaction between these modules. Much research is done in this direction (Bengio & Frasconi 1994; Cacciatore & Nowlan 1994; Jordan & Jacobs 1994; Rahman & Fairhurst 1999; Ronco, Gollee, & Gawthrop 1997). The alternative we want to propose here is to introduce a neural model that is capable of representing systems that are functionally equivalent to multi-expert systems within a single integrative network. This network does not explicitly distinguish between expert and gating modules and generalizes conventional neural networks by introducing a counterpart for gating interactions. What will be novel features of this new representation of multi-expert systems?

- First, our representation allows much more and qualitatively new architectural freedom. E.g., gating neurons may interact with expert neurons; gating neurons can be a part of experts. There is no restriction with respect to serial, parallel, or hierarchical architectures—in a much more general sense than proposed in (Jordan & Jacobs 1994).

- Second, our representation allows in an intuitive way to combine techniques from various learning theories. This includes gradient descent, unsupervised learning methods like Hebb learning or the Oja rule, and an EM-algorithm that can be transferred from classical gating-learning theories (Jordan & Jacobs 1994). Further, the interpretation of a specific gating as an action exploits the realm of reinforcement learning, in particular Q-learning and (though not discussed here) its $TD(\lambda)$ variants (Sutton & Barto 1998).

- Third, our representation makes a simple genetic encoding of such architectures possible. There already exist various techniques for evolutionary architecture optimization of networks (see Yao 1999 for a review). Applied on our representation, they become techniques for the evolution of multi-expert architectures.

Conventional multi-expert systems.   Assume the system has to realize a mapping from an input space X to an output space Y. Typically, an m-expert architecture consists of a gating function $\hat{g} : X \to [0,1]^m$ and m expert functions $f_i : X \to Y$ which are combined by the softmax linear combination:

$$y = \sum_{i=1}^{m} g_i f_i(x) , \quad g_i = \frac{e^{\beta \hat{g}_i(x)}}{\sum_{j=1}^{m} e^{\beta \hat{g}_j(x)}} , \tag{2.7}$$

where x and y are input and output, and β describes the "softness" of this winner-takes-all type competition between the experts, see figure 2.6. The crucial question becomes how to train the gating. We will discuss different methods below.

Neural implementation of multi-experts.   We present a single neural system that has at least the capabilities of a multi-expert architecture of several neural networks. Basically we provide additional competitive and gating interactions (for an illustration compare figure 2.6 and figure 2.7-B). We introduce the model as follows:

The architecture is given by a directed, labeled graph of neurons (i) and links (ij) from (j) to (i), where $i, j = 1..n$. Labels of links declare if they are ordinary, competitive, or gating connections. Labels of neurons declare their type of activation function. With every neuron (i), an activation state (output value) $z_i \in [0,1]$ is associated. A neuron (i) collects two terms of excitation x_i and g_i given by

$$x_i = \sum_{(ij)} w_{ij} z_j + w_i \tag{2.8}$$

$$g_i = \begin{cases} 1 & \text{if } N_i = 0 \\ \frac{1}{N_i} \sum_{(ij)^g} z_j & \text{otherwise} \end{cases} , \quad N_i = \sum_{(ij)^g} 1 , \tag{2.9}$$

where $w_{ij}, w_i \in \mathbb{R}$ are weights and bias associated with the links (ij) and the neuron (i), respectively. The second excitatory term g_i has the meaning of a gating term and is induced by N_i g-labeled links $(ij)^g$.

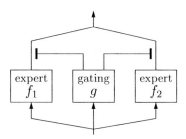

Figure 2.6: Ordinary multi-expert architecture. Gating and experts modules are explicitly separated and the gating may not depend on internal states or the output of experts.

In case there are no c-labeled links $(ij)^c$ connected to a neuron (i), its state is given by

$$z_i = \phi(x_i)\, g_i \ . \tag{2.10}$$

Here, $\phi : \mathbb{R} \to [0,1]$ is a sigmoid function. This means that, if a neuron (i) has no gating links $(ij)^g$ connected to it, then $g_i = 1$ and the sigmoid $\phi(x_i)$ describes its activation. Otherwise, the gating term g_i is multiplied.

Neurons (i) that are connected by (bi-directional) c-labeled links $(ij)^c$ form a *competitive group* in which only one of the neurons (the *winner*) acquires state $z_{\text{winner}} = 1$ while the other's states are zero. Let $\{i\}^c$ denote the competitive group of neurons to which (i) belongs. On this group, we introduce a normalized distribution y_i, $\sum_{j \in \{i\}^c} y_j = 1$, given by

$$y_i = \frac{\psi(x_i)}{X_i} \ , \quad X_i = \sum_{k \in \{i\}^c} \psi(x_k) \ . \tag{2.11}$$

Here, ψ is some function $\mathbb{R} \to \mathbb{R}$ (e.g., the exponential $\psi(x) = e^{\beta x}$). The neuron states $z_j \in \{0, 1\}$, $j \in \{i\}^c$ depend on this distribution y_i by one of the following competitive rules of winner selection: We will consider a selection with probability proportional to y_i (softmax), deterministic selection of the maximum y_i, and ϵ-greedy selection (where with probability ϵ a random winner is selected instead of the maximum).

Please see figure 2.7 to get an impression of the architectural possibilities this representations provides. Example A realizes an ordinary feed-forward neural

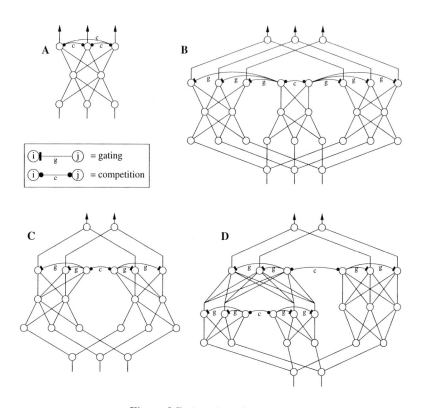

Figure 2.7: Sample architectures.

network, where the three output neurons form a competitive group. Thus, only one of the output neurons will return a value of 1, the others will return 0. Example B realizes exactly the same multi-expert system as depicted in figure 2.6. The two outputs of the central module form a competitive group and gate the output neurons of the left and right module respectively—the central module calculates the gating whereas the left and right modules are the experts. Example C is an alternative way of designing multi-expert systems. Each expert module contains an additional output node which gates the rest of its outputs and competes with the gating nodes of the other experts. Thus, each expert estimates itself how good it can handle the current stimulus (see the Q-learning method described below). Finally, example D is a true hierarchical architecture. The two experts on the left compete to give an output, which is further processed and, again, has to compete with the larger expert to the right. In contrast, Jordan & Jacobs (1994) describe an architecture where the calculation of one single gating (corresponding to only one competitive level) is organized in a hierarchical manner. Here, several gatings on different levels can be combined in any successive, hierarchical way.

In the following we introduce four different learning methods, each of which is applicable independently of the specific architecture. We generally assume that the goal is to approximate training data given as pairs (x, t) of stimulus and target output value.

2.3.1.1   Gradient learning

To calculate the gradient, we assume that selection in competitive groups is performed with probability proportional to the distribution y_i. We calculate an approximate gradient of the conditional probability $\mathcal{P}(y|x)$ that this system represents by replacing the actual state z_i in eq. (2.8) by its expectation value y_i for neurons in competitive groups (see also Neal 1990). For simplicity of notation, we identify $z_i \equiv y_i$. Then, for a neuron (i) in a competitive group obeying eq. (2.11), we get the partial derivatives of the neuron's output with respect to its excitations:

$$\frac{\partial z_i}{\partial x_j} = \frac{\psi'(x_i)\,\delta_{ij}}{X_i} - \frac{\psi(x_i)}{(X_i)^2}\left[\psi'(x_j)\,\delta_{j\in\{i\}^c}\right]$$

$$= \frac{\psi'(x_j)}{X_i}\left[\delta_{ij} - z_i\,\delta_{j\in\{i\}^c}\right], \tag{2.12}$$

$$\frac{\partial z_i}{\partial g_j} = 0 , \tag{2.13}$$

where $\delta_{j\in\{i\}^c} = 1$ iff j is a member of $\{i\}^c$. Let $E = E(z_1, .., z_n)$ be an error functional. We write the delta-rule for back-propagation by using the notations $\breve{\delta}_i = \frac{dE}{dz_i}$, $\delta_i = \frac{dE}{dx_i}$, and $\delta_i^g = \frac{dE}{dg_i}$ for the gradients at a neuron's output and inputs, respectively, and $e_i = \frac{\partial E}{\partial z_i}$ for the local error of a single (output) neuron. From eqs. (2.8, 2.9, 2.12, 2.13) we get

$$\breve{\delta}_i = \frac{dE}{dz_i} = e_i + \sum_j \frac{dE}{dx_j}\frac{\partial x_j}{\partial z_i} + \sum_j \frac{dE}{dg_j}\frac{\partial g_j}{\partial z_i}$$

$$= e_i + \sum_{(ji)} \delta_j \, w_{ji} + \sum_{(ji)^g} \delta_j^g \, \frac{1}{N_i} \ , \tag{2.14}$$

$$\delta_i = \frac{dE}{dx_i} = \sum_j \check{\delta}_j \, \frac{\partial z_j}{\partial x_i} = \frac{\psi'(x_i)}{X_i} \, \big[\check{\delta}_i - \sum_{j \in \{i\}^c} \check{\delta}_j \, z_j \big] \ , \tag{2.15}$$

$$\delta_i^g = \frac{dE}{dg_i} = \sum_j \check{\delta}_j \, \frac{\partial z_j}{\partial g_i} = 0 \ . \tag{2.16}$$

(In eq. (2.15) we used $X_i = X_j$ for $i \in \{j\}^c$ and $i \in \{j\}^c \Leftrightarrow j \in \{i\}^c$.) For neurons that do not join a competitive group we get from eq. (2.10)

$$\frac{\partial z_i}{\partial x_j} = \phi'(x_i) \, g_i \, \delta_{ij} \ , \qquad \frac{\partial z_i}{\partial g_j} = \phi(x_i) \, \delta_{ij} \ , \tag{2.17}$$

$$\delta_i = \frac{dE}{dx_i} = \sum_j \check{\delta}_j \, \frac{\partial z_i}{\partial x_j} = \phi'(x_i) \, g_i \, \check{\delta}_i \ , \tag{2.18}$$

$$\delta_i^g = \frac{dE}{dg_i} = \sum_j \check{\delta}_j \, \frac{\partial z_i}{\partial g_j} = \phi(x_i) \, \check{\delta}_i \ , \tag{2.19}$$

where $\check{\delta}_i$ is given in eq. (2.14). The final gradients are

$$\frac{dE}{dw_i} = \delta_i \ , \qquad \frac{dE}{dw_{ij}} = \delta_i \, z_j \ . \tag{2.20}$$

The choice of the error functional is arbitrary. E.g., it can be chosen as the square error $E = \sum_i (z_i - t_i)^2$, $e_i = 2(z_i - t_i)$ or as the negative log-likelihood $E = -\ln \prod_i z_i^{t_i} (1 - z_i)^{t_i}$, $e_i = \frac{1-t_i}{1-z_i} - \frac{t_i}{z_i}$, where in the latter case the target are states $t_i \in \{0, 1\}$.

The basis for further learning rules.   For the following learning methods we concentrate on the question: *What target values should we assume for the states of neurons in a competitive group?* In the case of gradient descent, eq. (2.14) gives the answer. It actually describes a linear projection of the desired output variance down to all system states z_i—including those in competitions. In fact, all the following learning methods will adopt the above gradient descent rules except for a redefinition of $\check{\delta}_i$ (or alternatively δ_i) in the case of neurons (i) in competitive groups. This means that neurons "below" competitive groups are adapted by ordinary gradient descent while the local error *at* competitive neurons is given by other rules than gradient descent. Actually this is the usual way for adapting systems where neural networks are used as internal modules and trained by back-propagation (see, e.g., Anderson & Hong 1994).

2.3.1.2 EM-learning

We briefly review the basic ideas of applying an EM-algorithm to the problem of learning gatings in multi-experts (Jordan & Jacobs 1994). The algorithm is based on an additional, very interesting assumption: Let the outcome of a competition in a competitive group $\{c\}^c$ be described by the states $z_i \in \{0, 1\}$, $\sum_{i \in \{c\}^c} z_i = 1$ of the neurons that are members of this group. Now, we assume that there exists a *correct* outcome $h_i \in \{0, 1\}$, $\sum_{i \in \{c\}^c} h_i = 1$. Formally, this means to assume that the complete training data are triplets (x, h_i, t) of stimuli, competition states, and output values.[1] However, the competition training data is unobservable or *hidden* and must be inferred by statistical means. Bayes' rule gives an answer on how to infer an expectation of the hidden training data h_i and lays the ground for an EM-algorithm. The consequence of this assumption is that now the y_i of competitive neurons are supposed to approximate this expectation of the training data h_i instead of being free. For simplification, let us concentrate on a network containing a single competitive group; the generalization is straightforward.

- Our system represents the conditional probability of output states z^o and competition states z^c, depending on the stimulus x and parameters $\theta = (w_{ij}, w_i)$:

$$\mathcal{P}(z^o, z^c | x, \theta) = \mathcal{P}(z^c | x, \theta)\, \mathcal{P}(z^o | z^c, x, \theta) . \tag{2.21}$$

- (E-step) We use Bayes rule to infer the expected competition training data h_i hidden in a training tuple $(x, \cdot, t)$, i.e., the probability of h_i when x and t are given.

$$\mathcal{P}(h_i | x, t) = \frac{\mathcal{P}(t | h_i, x) \mathcal{P}(h_i | x)}{\mathcal{P}(t | x)} \tag{2.22}$$

---

[1]More precisely, the assumption is that there exists a teacher system of same architecture as our system. Our system adapts free parameters w_{ij}, w_i in order to approximate this teacher system. The teacher system produces training data and, since it has the same architecture as ours, also uses competitive groups to generate this data. The training data would be complete if it included the outcomes of these competitions.

Since these probabilities refer to the training (or teacher) system, we can only approximate them. We do this by our current approximation, i.e., our current system:

$$\mathcal{P}(h_i|x,t,\theta) = \frac{\mathcal{P}(t|h_i,x,\theta)\,\mathcal{P}(h_i|x,\theta)}{\mathcal{P}(t|x,\theta)}$$

$$= \frac{\mathcal{P}(t|h_i,x,\theta)\,\mathcal{P}(h_i|x,\theta)}{\sum_{z^c}\mathcal{P}(z^c|x,\theta)\,\mathcal{P}(t|z^c,x,\theta)} \ . \tag{2.23}$$

- (M-step) We can now adapt our system. In the classical EM-algorithm, this amounts to maximizing the expectation of the log-likelihood (cp. eq. (2.21))

$$\mathrm{E}[l(\theta')] = \mathrm{E}[\ln \mathcal{P}(h|x,\theta') + \ln \mathcal{P}(t|z^c,x,\theta')] \ , \tag{2.24}$$

where the expectation is with respect to the distribution $\mathcal{P}(h|x,t,\theta)$ of h-values (i.e., depending on our inference of the hidden states h); and the maximization is with respect to parameters θ. This equation can be simplified further—but, very similar to the "least-square" algorithm developed by Jordan & Jacobs (1994), we are satisfied to have inferred an explicit *desired* probability $\hat{y}_i = \mathcal{P}(h_i = 1|x,t,\theta)$ for the competition states z_i that we use to define a mean square error and perform an ordinary gradient descent.

Based on this background we define the learning rule as follows and with some subtle differences to the one presented by (Jordan & Jacobs). Equation (2.23) defines the desired probability $\hat{y}_i$ of the states z_i. Since we assume a selection rule proportional to the distribution y_i, the values $\hat{y}_i$ are actually target values for the distribution y_i. The first modification we propose is to replace all likelihood measures involved in eq. (2.23) by general error measures E: Let us define

$$Q_i(x) := 1 - E(x) \quad \text{if } (i) \text{ wins.} \tag{2.25}$$

Then, in the case of the likelihood error $E(x) = 1 - \mathcal{P}(t|x,\theta)$, we retrieve $Q_i(x) = \mathcal{P}(t|h_i = 1, x, \theta)$. Further, let

$$V(x) := \sum_i Q_i(x)\,y_i(x). \tag{2.26}$$

By these definitions we may rewrite eq. (2.23) as

$$\hat{y}_i(x) = \frac{Q_i(x)\,y_i(x)}{V(x)} = \frac{Q_i(x)\,y_i(x)}{\sum_j Q_j(x)\,y_j(x)}\;. \tag{2.27}$$

However, this equation needs some discussion with respect to its explicit calculation in our context—leading to the second modification. Calculating $Q_j(x)$ for every j amounts to evaluating the system for every possible competition outcome. One major difference to the algorithm presented in (Jordan & Jacobs 1994) is that we do not allow for such a separate evaluation of all experts in a single time step. In fact, this would be very expensive in case of hierarchically interacting competitions and experts because the network had to be evaluated for each possible combinatorial state of competition outcomes. Thus we propose to use an approximation: We replace $Q_j(x)$ by its average over the recent history of cases where (j) won the competition,

$$\bar{Q}_j \leftarrow \gamma\,\bar{Q}_j + (1-\gamma)\,Q_j(x) \quad \text{whenever } (j) \text{ wins}\,, \tag{2.28}$$

where $\gamma \in [0, 1]$ is a trace constant (as a simplification of the time dependent notation, we use the algorithmic notation $\leftarrow$ for a replacement if and only if (j) wins). Hence, our adaptation rule finally reads

$$\breve{\delta}_i = -\alpha_c \left[y_i - \frac{Q_i\,y_i}{\sum_{j\in\{i\}^c}\bar{Q}_j\,y_j} \right] \quad \text{if } (i) \text{ wins}, \tag{2.29}$$

and $\breve{\delta}_i = 0$ if (i) does not win; which means a gradient descent on the square error between the approximated desired probabilities $\hat{y}_i$ and the distribution y_i.

2.3.1.3 Q-learning

Probably, the reader has noticed that we chose notations in the previous section in the style of reinforcement learning: If one interprets the winning of neuron (i) as a decision on an action, then $Q_i(x)$ (called *action-value function*) describes the (estimated) quality of taking this decision for stimulus x; whereas $V(x)$ (called *state-value function*) describes the estimated quality for stimulus x without having decided yet, see (Sutton & Barto). In this context, eq. (2.27) is very interesting: It proposes to adapt the probability $y_i(x)$ according to the ratio $Q_i(x)/V(x)$—the EM-algorithm acquires a very intuitive interpretation. To

– The adaptation rate is $\alpha = 0.01$ for all algorithms (as indicated in eqs. (2.29,2.30,2.32), the delta-values for neurons in competitive groups are multiplied by the learning rate α_c).

– Parameters are initialized normally distributed around zero with standard deviation $\sigma = 0.01$.

– The sigmoidal and linear activation functions are $\phi_s(x) = \frac{1}{1+\exp(-10x)}$ and $\phi_l(x) = x$, respectively.

– The competition function ψ for softmax competition is $\psi_s(x) = e^{5x}$.

– The Q-learning algorithm uses ϵ-greedy selection with $\epsilon = 0.1$; the others select either the maximal activation or with probability proportional to the activation.

– The values of the average traces $\bar{Q}_i$ and $\bar{V}$ are initialized to 1.

– The following parameters were used for the different learning schemes:

	gradient	EM	Q	Oja-Q
α_c	–	1	10	100
γ	–	0.9	–	0.9
ψ	ψ_s	ϕ_s	ϕ_l	ϕ_l
selection	proportional	max	greedy	max

Here, α_c is the learning rate factor, γ is the average trace parameter, and ψ is the competition function.

Table 2.2: Implementation details.

realize this equation without the approximation described above one has to provide an estimation of $V(x)$, e.g., a neuron trained on this target value (a *critic*). We leave this for future research and instead directly address the Q-learning paradigm.

For Q-learning, an explicit estimation of the action-values $Q_i(x)$ is modeled. In our case, we realize this by considering $Q_i(x)$ as the target value of the excitations x_i, $i \in \{c\}^c$, i.e., we train the excitations of competing neurons toward the action values,

$$\delta_i = \alpha_c \begin{cases} x_i - Q_i & \text{if } (i) \text{ wins} \\ 0 & \text{else .} \end{cases} \quad (2.30)$$

This approach seems very promising—in particular, it opens the door, e.g., to TD(λ) methods and other fundamental concepts of reinforcement learning theory.

2.3.1.4 Oja-Q learning

Besides statistical and reinforcement learning theories, also the branch of unsupervised learning theories gives some inspiration for our problem. The idea of hierarchically, serially coupled competitive groups raises a conceptual problem: Can competitions in areas close to the input be trained while higher level areas (closer to the output) are not yet operative and vice versa? Usually, backpropagation is the standard technique to address this problem. But this does not apply on either the EM-learning or the reinforcement learning approaches because they generate a direct feedback to competing neurons in any layer. Unsupervised learning in lower areas seems to show a way out of this dilemma. As a first approach we propose a mixture of unsupervised learning in the fashion of the normalized Hebb rule and Q-learning. The normalized Hebb rule (of which the Oja rule is a linearized version) can be realized by setting $\delta_i = -\alpha_c z_i$ for a neuron (i) in a competitive group (recall $z_i \in \{0, 1\}$). The gradient descent with respect to adjacent input links gives the ordinary $\Delta w_{ij} \propto z_i z_j$ rule. Thereafter, the input weights (including the bias) of each neuron (i), $i \in \{c\}^c$ are normalized. We modify this rule in two respects. First, we introduce a factor $(Q_i - \bar{V})$ that accounts for the success of neuron (i) being the winner. Here, $\bar{V}$ is an average trace of the feedback:

$$\bar{V} \leftarrow \gamma \bar{V} + (1 - \gamma) Q_i(x) \quad \text{every time step} , \tag{2.31}$$

where (i) is the winner. Second, in the case of failure, $Q_i < \bar{V}$, we also adapt the non-winners in order to increase their response on the stimulus next time. Thus, our rule reads

$$\delta_i = -\alpha_c (Q_i - \bar{V}) \begin{cases} z_i & \text{if } Q_i \geq \bar{V} \\ z_i - 0.5 & \text{else} . \end{cases} \tag{2.32}$$

Similar modifications are often proposed in reinforcement learning models (Barto 1985; Barto & Jordan 1987). The rule investigated here is only a first proposal; all rules presented in the excellent survey of Diamantaras & Kung (1996) can equally be applied.

2.3.2 Empirical tests

Some tests of the proposed architecture and the learning schemes on a task similar to the what-and-where task demonstrate the functionality.

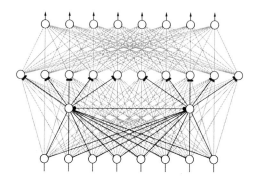

Figure 2.8: The architecture we use for our experiments. All output neurons have linear activation functions $\phi(x) = x$. All neurons except the input neurons have bias terms.

The task. We test the functionality of our model and the learning methods by addressing a variant of the *what-and-where* task also investigated by (Jacobs, Jordan, & Barto). A single bit of an 8-bit input decides on the subtask that the system has to solve on the current input. The two subtasks itself are rather simple and in our case (other than in Jacobs, Jordan, & Barto 1990) are to map the 8-bit input either identically or inverted on the 8-bit output. The task has to be learned online.

The architectures. We investigate the learning dynamics of our model with the 4 different learning methods. We use a fixed architecture similar to an 8-10-8-layered network with 10 hidden neurons but additionally install 2 competitive neurons that receive the input, each of which gates half of the hidden neurons, see figure 2.8. For completeness we also display the learning curve of a conventional feed-forward neural network (FFNN) in which case we used the same architecture but replaced all gating and competitive connections by conventional links.

Results. Figure 2.9 displays the learning curves for each case averaged over 20 runs with different weight initializations. For implementation details see table 2.2. First of all, we find that all of the 4 learning methods seem to work and

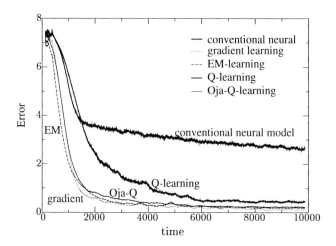

Figure 2.9: Learning curves for the conventional neural network and the four different learning schemes averaged over 20 runs with different weight initializations.

perform well on this task. Not surprisingly, the conventional FFNN fails. It was only *sometimes* able to solve the task completely what explains the rather high error offset for its learning curve.

We get some more insight in the learning dynamics by investigating if and when the task separation has been learned. Figure 2.10 displays the frequencies of winning of the two competitive neurons in case of the different subtasks: "How often does the left or right neuron win in case of the first and the second task?" The task separation would be perfect if these two neurons would reliably distinguish the two subtasks. First noticeable is that all 4 learning methods learn the task separation. Second, after the task separation has been found in principle, more or less noise is left, depending on learning scheme. However, this second effect only mirrors the respective "action-selection policy": For Q-learning we used ϵ-greedy selection of the winner with $\epsilon = 0.1$. Hence, the actually perfectly learned task separation is distorted with 10% noise. This is similar with the pure gradient method, where softmax selection introduces additional noise. Only EM and Oja-Q display the perfect task separation since they use maximum winner selection.

The four learning methods only differ in how the two gating neurons are trained.

Consequently, if the gating neurons solved their problem of task-separation, the following epoch of learning the two separate tasks is equivalent for all four learning methods. This explains why the four learning curves in figure 2.9 are so similar except for the temporal offset corresponding to the time until the task separation has been found and the non-zero asymptotic error corresponding to the noise of task separation.

Generally, our experience was that the learning curves may look very different depending on the weight initialization. It also happened that the task separation was not found when weights and biases (especially of the competing neurons) were initialized with very large values (e.g., by $\mathcal{N}(0, 0.5)$). One of the competitive neurons then dominates from the very beginning and prohibits the "other expert" to adapt in any way. Definitely, a special, perhaps equal initialization of competitive neurons could be profitable.

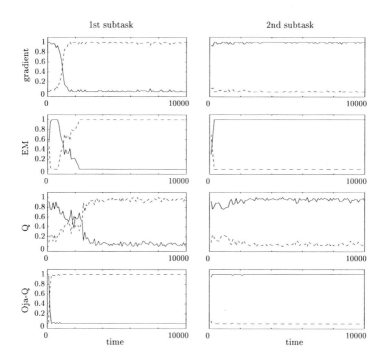

Figure 2.10: The gating ratios for single trials for the four different learning schemes: The four rows refer to gradient, EM-, Q-, and Oja-Q-learning; and the two columns refer to the two classes of stimuli—one for the "identical" task, and one for the "not" task. Each graph displays two curves that sum to 1 and indicate how often the first or second gating neuron wins in case of the respective subtask.

Conclusions

A key result of our discussion of evolutionary adaptation was that evolution implicitly learns about the problem by adapting its genetic representations accordingly. The emerging genetic representations, including their neutral traits, encode knowledge about where to explore. The information is accumulated in the course of σ-evolution of the genetic systems. This theory together with our experimental studies allows to understand some fundamental phenomena of evolution:

- Evolution can adapt genetic representations in order to induce *correlated* phenotypic variability as we observed it for the evolution of artificial plants in section 1.5.5. The origin of the structuredness of phenotypic variability are basic correlating mechanisms within the genotype-phenotype mapping (e.g., gene interaction mechanisms like the operon) that are exploited by the adaptation of genetic representations.

- Evolution adapts genetic representations towards shorter and more modular genomes if otherwise phenotypic mutability becomes too high. In some cases the price is that optimal solutions are not found since they would require mutationally non-stable encodings, as, e.g., in the example in section 1.5.4.

- Neutrality is not redundant in the sense of superfluous since different genetic representations within a neutral set induce different phenotypic variability. Hence, neutral traits carry information measurable in terms of phenotypic exploration distributions.

- Evolution exploits neutrality when adapting genetic representations with-

out affecting the current phenotype. In our comput.   ona   del, structural mutations of the genome—2nd-type mutation: acc   for such
exploitation of neutrality and enable structural refo   tio   f genetic
representations.

- Redundancy in the sense of multiple genes for the s   e p   se may
  occur when evolution adapts genetic representations in f   r of   notypic
  mutational robustness. For instance, this occurs in the   rial   section
  1.5.5 when several operators with the same promoter exi: f w   some
  would only be expressed when the other is mutationally d roy

We claim that all these phenomena are merely different faces of   un   lying
principle of σ-evolution. This is the adaptation of genetic represen   ions   iven
by the pressure to match phenotypic variability to the distribution   xpe   ced
good solutions—the fundamental principle to learn from previous   plo   ns
by accumulating the information and exploiting it for future explo   ion   we
captured it formally in terms of Generic Heuristic Search.

In the introduction we formulated the idea of "simple adaptation   hani   s
on suitable representations" versus "complex adaptation mechanism   n a
trary representations." Our results show that indeed the adaptation   ;enc
representations allows that a very simple adaptational mechanism on   1e   e
tual substrate of evolutionary adaptation—on the level of genes—ma   ,du
elaborate adaptation on the functional level.

The adaptability of neural systems can be understood and characterized   basically the same way by analyzing how system functionalities are represent   by
the underlying adaptive substrate—the synaptic (and bias) weights. We   el-
oped a theory on how the structure of neural systems determines their sty   f
functional adaptability. Unlike classical approaches to the model selection p.
lem, which focus on the capacity of the system (i.e., the cardinality of the sp.
of representable functions, e.g., the VC-dimension), our analysis captures prec.
characteristics of *how* functions are represented and what style of adaptabili
characterized by coadaptation, this induces. This description of how function
are represented leads to a definition of *functional* modularity which contrasts
to existing definitions of architectural modularity in neural systems (see, e.g.,
Hüsken, Igel, & Toussaint 2002) which do not a priory have a functional meaning.

As an example for the significance of this theory we analyzed two types of neural
systems; conventional neural networks with and without lateral competitive in-

teractions. This analysis makes explicit and theoretically precise what has often been discussed:

- Despite the universal approximation capabilities of neural networks, their way of adaptability, in the case of gradient learning, is not universal: They are inherently non-modular and predisposed for coadaptation.

- Competitive interactions, as they are readily implicit in multi-expert systems to solve the problem of cross-talk, can prevent undesired coadaptation and induce a much more modular way of adaptability in the sense of definition 2.2.2.

Both results become apparent when investigating the distribution of the functional metric over the respective model class, see figure 2.5. Such distributions are a generic way to characterize model classes with respect to their adaptation behavior.

These results have important implications for the design of neural systems—in particular for the approach to use evolutionary algorithms to optimize the architecture of neural systems: If one is interested in optimizing the way of adaptability and the way of generalization of neural systems, one should not only consider the space of conventional neural networks as the search space but also include systems that incorporate competitive interactions. We defined such a general class of neural networks that may incorporate competitive interactions as well as gating interactions anywhere in the architecture and thereby unifies and generalizes the class of conventional neural networks and the class of general architecture multi-expert systems. Evolutionary architecture optimization within this class of networks should lead to neural systems of which functional adaptability is well adapted to the problem at hand. E.g., functional traits that are decoupled in the problem should be represented by different functional modules of the neural system.

Appendix

Theorems, definitions and symbols in chapter 1

symbol	page (eq.)	description
P	24, 32	space of solutions, the search space (in the context of heuristic search) or the phenotype space
Λ^P	24	space of distributions over the search space; space of phenotypic exploration distributions
$q^{(t)} = \Phi y^{(t)} \in \Lambda^P$	24	exploration distribution at time t
$y^{(t)} \in Y$	24	parameters of the exploration distribution at time t
$\Phi : Y \to \Lambda^P$	24	parameterization of the exploration distribution
$S^\lambda : \Lambda^P \to \Lambda^P$	25	stochastic sampling operator makes a distribution (population) finite
$\mathcal{F} : \Lambda^P \to \Lambda^P$	25 (1.1)	fitness operator
$\mathcal{H}$	25 (1.2)	heuristic rule (essentially mapping $y^{(t)}$ to new parameters $y^{(t+1)}$)
$D(p : q)$	26 (1.3)	Kullback-Leibler divergence ("distance measure" between two distributions)
G	28, 33	genotype space
Λ^G	28	space of distributions over genotype space, i.e., the space of populations also finite populations
$p^{(t)} \in \Lambda^G$	28	genotype population
$\mathcal{M} : \Lambda^G \to \Lambda^G$	29	mutation or mixing (mutation plus crossover) operator
$q^{(t)} = \mathcal{M} p^{(t)} \in \Lambda^G$	29	genotype offspring population $q^{(t)} = \mathcal{M} p^{(t)}$ (which is at the same time the genotypic exploration distribution)
$\Lambda^{\mu, G}$	31	space of *finite* distributions (i.e., populations) comprising μ samples (individuals)
$\phi : G \to P$	33	genotype phenotype mapping
$\equiv$	33	phenotype equivalence
$[x] \subseteq G$	33	neural set of the phenotype $x \in P$ (i.e., the equivalence class of genotypes with the same phenotype x)
$\Xi : \Lambda^G \to \Lambda^P$	33	phenotype projection of distributions (populations)
$\hateq$	33	phenotype equivalence between distributions (populations) of genotypes
$n(g)$	34	neutral degree of a genotype $g \in G$
$\overline{[x]} \subset \Lambda^G$	42 (1.9)	set of exploration distributions in a neutral set $[x]$
$g \mapsto (x, \sigma)$	42	typical nomenclature of the σ-embedding: genotype g, phenotype $x = \phi(g)$, and the "neutral traits" $\sigma = \mathcal{M}(\cdot|g) \in \Lambda^G$ embedded in the space of distributions
$\mathcal{C} : \Lambda^G \to \Lambda^G$	52	crossover operator
$\mathcal{M}^*$	52	a simple mutation operator (the typical componentwise (i.e., symbol-wise) mutation)

Theorems, definitions and symbols in chapter 2

symbol	page (eq.)	description
F	110	functional space of neural systems, e.g., the space of functions or conditional probability distributions
W	111	parameter space of neural systems, i.e., the weight space $\mathbb{R}^m$
Λ^W, Λ^F	111	space of distributions over parameter and functional spaces, respectively
$\Phi : W \to F_d$	111, 115, 118	parameterization of functionalities
$w \in W = \mathbb{R}^m$	115	parameter state of a neural system
$\mathcal{H} : W \to \Lambda^W$	115	stochastic adaptation operator, mapping $w^{(t)}$ to a probability distribution for $w^{(t+1)}$
$\mathcal{S}^\lambda : \Lambda^W \to W$	115	stochastic sampling operator; see the definition on page 25 in chapter 1
$f \in F$	116	functionality (function) represented by the neural system
$f^a \in \mathbb{R}$	116	functional components
$C_{ab}(w)$	116	covariance matrix between functional traits within stochastic adaptation
$\langle \cdot \rangle$	116	averaging over a specified distribution
$\Xi : \Lambda^W \to \Lambda^F$	117	projection of distributions from parameter space to the functional space; see definition 33 (1.2.8) in chapter 1
$\hat{g}_{ij}$	118, 120	an ad-hoc presumed metric on the weight space W, namely the Euclidean metric
$g_{ab}(w)$	119	induced functional metric on F
g_{ab}	122	a presumed (natural) metric on F (namely the mean square metric or the Fisher metric)
$g_{ij}(w)$	122	weight space metric induced from the natural metric $\hat{g}_{ab}$ in the case of natural gradient descent
δ_{ij}	120	Euclidean metric or, equivalently, the Kronecker delta
α	119	adaptation rate
$\boldsymbol{\xi}_a$	119	local basis vectors of the coordinate frame in F

References

Akaike, H. (1974). A new look at the statistical model identification. *IEEE Transactions on Automatic Control* **AC–19**, 716–723. For a reprint see E. Parzen et al. (Eds.), *Selected Papers of Hirotugu Akaike*, Springer Series in Statistics, 1998.

Altenberg, L. (1994). Evolving better representations through selective genome growth. In *Proc. of IEEE World Congress on Computational Intelligence*, pp. 182–187.

Altenberg, L. (1995). Genome growth and the evolution of the genotype-phenotype map. In W. Banzhaf & F. H. Eeckman (Eds.), *Evolution and Biocomputation: Computational Models of Evolution*, pp. 205–259. Springer, Berlin.

Amari, S. (1993). Mathematical methods of neurocomputing. In O. Barndorff-Nielsen, J. Jensen, & W. Kendall (Eds.), *Networks and Chaos—Statistical and Probabilistic Aspects*, pp. 1–39. Chapman & Hall, London.

Amari, S. (1998). Natural gradient works efficiently in learning. *Neural Computation* **10**, 251–276.

Amari, S., H. Park, & K. Fukumizu (2000). Adaptive method of realizing natural gradient learning for multilayer perceptrons. *Neural Computation* **12**, 1399–1409.

Anderson, C. W. & Z. Hong (1994). Reinforcement learning with modular neural networks for control. In *IEEE Int. Workshop on Neural Networks Application to Control and Image Processing*.

Angeline, P. (1995). Adaptive and self-adaptive evolutionary computations. In M. Palaniswami, Y. Attikiouzel, R. Marks, D. B. Fogel, & T. Fukuda (Eds.), *Computational Intelligence: A Dynamic Systems Perspective*, pp. 152–163. IEEE Press.

Bäck, T. (1996). *Evolutionary Algorithms in Theory and Practice*. Oxford University Press.

Bäck, T. (1998a). On the behavior of evolutionary algorithms in dynamic environments. In D. Fogel, H.-P. Schwefel, T. Bäck, & X. Yao (Eds.), *Proc. of Fifth IEEE Int. Conf. on Evolutionary Computation (ICEC 1998)*, pp. 446–451. IEEE Press.

Bäck, T. (1998b). An overview of parameter control methods by self-adaptation in evolutionary algorithms. *Fundamenta Informaticae* **35**, 51–66.

Baluja, S. (1994). Population-based incremental learning: A method for integrating genetic search based function optimization and competitive learning. Technical Report CMU-CS-94-163, Comp. Sci. Dep., Carnegie Mellon U.

Baluja, S. & S. Davies (1997). Using optimal dependency-trees for combinatorial optimization: Learning the structure of the search space. In *Proc. of Fourteenth Int. Conf. on Machine Learning (ICML 1997)*, pp. 30–38.

Barto, A. (1985). Learning by statistical cooperation of self-interested neuron-like computing elements. *Human Neurobiology* **4**, 229–256.

Barto, A. & M. Jordan (1987). Gradient following without back-propagation in layered networks. In *Proc. of IEEE First Annual Conf. on Neural Networks*, pp. II629–II636.

Bengio, Y. & P. Frasconi (1994). An EM approach to learning sequential behavior. Technical Report DSI-11/94, Dipartimento di Sistemi e Informatica, Universita di Firenze.

Beyer, H.-G. (2001). *The Theory of Evolution Strategies*. Springer, Berlin.

Cacciatore, T. W. & S. J. Nowlan (1994). Mixtures of controllers for jump linear and non-linear plants. In J. D. Cowan, G. Tesauro, & J. Alspector (Eds.), *Advances in Neural Information Processing Systems*, Volume 6, pp. 719–726. Morgan Kaufmann.

Chenn, A. & C. A. Walsh (2002). Regulation of cerebral cortical size by control of cell cycle exit in neural precursors. *Science* **297**, 365–369.

Conrad, M. (1990). The geometry of evolution. *BioSystems* **24**, 61–81.

Cover, T. A. & J. A. Thomas (1991). *Information Theory*. John Wiley & Sons, New York.

Diamantaras, K. I. & S. Y. Kung (1996). *Principle component neural networks: theory and applications*. John Wiley & Sons, New York.

Eigen, M., J. McCaskill, & P. Schuster (1989). The molecular quasispecies. *Advances in chemical physics* **75**, 149–263.

Eigen, M. & P. Schuster (1977). The hypercycle. *Die Naturwissenschaften* **64**, 541–565.

Fontana, W. & P. Schuster (1998). Continuity in evolution: On the nature of transitions. *Science* **280**, 1431–1433.

French, R. M. (1999). Catastrophic forgetting in connectionist networks. *Trends in Cognitive Scienences* **3**, 128–135.

Gruau, F. (1995). Automatic definition of modular neural networks. *Adaptive Behaviour* **3**, 151–183.

Halder, G., P. Callaerts, & W. Gehring (1995). Induction of ectopic eyes by targeted expression of the eyeless gene in Drosophila. *Science* **267**, 1788–1792.

Hansen, N. & A. Ostermeier (2001). Completely derandomized self-adaption in evolutionary strategies. *Evolutionary Computation* **9**, 159–195.

Hansen, T. F. & G. P. Wagner (2001a). Epistasis and the mutation load: A measurement-theoretical approach. *Genetics* **158**, 477–485.

Hansen, T. F. & G. P. Wagner (2001b). Modeling genetic architecture: A multilinear model of gene interaction. *Theoretical Population Biology* **59**, 61–86.

Holland, J. (1975). *Adaptation in Natural and Artificial Systems*. University of Michigan Press, Ann Arbor, USA.

Holland, J. H. (2000). Building blocks, cohort genetic algorithms, and hyperplane-defined functions. *Evolutionary Computation* **8**, 373–391.

Hornby, G. S. & J. B. Pollack (2001a). The advantages of generative grammatical encodings for physical design. In *Proc. of 2001 Congress on Evolutionary Computation (CEC 2001)*, pp. 600–607. IEEE Press.

Hornby, G. S. & J. B. Pollack (2001b). Evolving L-systems to generate virtual creatures. *Computers and Graphics* **25**, 1041–1048.

Hornik, K., M. Stinchcombe, & H. White (1989). Multilayer feedforward networks are universal approximators. *Neural Networks* **2**, 359–366.

Hüsken, M., C. Igel, & M. Toussaint (2002). Task-dependent evolution of modularity in neural networks. *Connection Science* **14**, 219–229.

Igel, C. & M. Toussaint (2003a). Neutrality and Self-adaptation. *Natural Computation* **2**, 117–132.

Igel, C. & M. Toussaint (2003b). On classes of functions for which no free lunch results hold. *Information Processing Letters* **86**, 317–321.

Jacob, F. & J. Monod (1961). Genetic regulatory mechanisms in the synthesis of proteins. *Journal of Molecular Biology* **3**, 318–356.

Jacobs, R. (1999). Computational studies of the development of functionally specialized neural modules. *Trends in Cognitive Sciences* **3**, 31–38.

Jacobs, R. A., M. I. Jordan, & A. G. Barto (1990). Task decomposistion through competition in a modular connectionist architecture: The what and where vision tasks. Technical Report COINS-90-27, Department of Computer and Information Science, University of Massachusetts Amherst.

Jordan, M. I. & R. A. Jacobs (1994). Hierarchical mixtures of experts and the EM algorithm. *Neural Computation* **6**, 181–214.

Kauffman, S. (1989). Adaptation on rugged fitness landscapes. In D. Stein (Ed.), *lectures in the Sciences of Complexity*, pp. 527–618. Addison-Wesley, Redwood City.

Kearns, M., Y. Mansour, A. Y. Ng, & D. Ron (1995). An experimental and theoretical comparison of model selection methods. In *Proc. of Workshop on Computational Learning Theory (COLT 1995)*. Morgan Kaufmann.

Kimura, M. (1983). *The Neutral Theory of Molecular Evolution*. Cambridge University Press.

Kimura, M. (1986). DNA and the Neutral Theory. *Philosophical Transactions, Royal Society of London* **B312**, 343–354.

Kitano, H. (1990). Designing neural networks using genetic algorithms with graph generation systems. *Complex Systems* **4**, 461–476.

Knippers, R. (1997). *Molekulare Genetik* (7 ed.). Georg Thieme, Stuttgart.

Kullback, S. & R. Leibler (1951). On information and sufficiency. *Annals of Mathematical Statistics* **22**, 7986.

Lucas, S. (1995). Growing adaptive neural networks with graph grammars. In *Proc. of European Symp. on Artificial Neural Netw. (ESANN 1995)*, pp. 235–240.

Moody, J. (1991). The effective number of parameters: an analysis of generalization and regularization in nonlinear systems. In *Advances in Neural Information Processing Systems*, Volume 4, pp. 847–854.

Mühlenbein, H., T. Mahnig, & A. O. Rodriguez (1999). Schemata, distributions and graphical models in evolutionary optimization. *J. of Heuristics* **5**, 215–247.

Neal, R. M. (1990). Learning stochastic feedforward networks. Technical Report CRG-TR-90-7, Department of Computer Science, University of Toronto.

Nordin, P. & W. Banzhaf (1995, 15-19). Complexity compression and evolution. In L. Eshelman (Ed.), *Genetic Algorithms: Proc. of Sixth International Conf. (ICGA 1995)*, pp. 310–317. Morgan Kaufmann, Pittsburgh.

Pelikan, M., D. E. Goldberg, & E. Cantú-Paz (2000). Linkage problem, distribution estimation, and Bayesian networks. *Evolutionary Computation* **9**, 311–340.

Pelikan, M., D. E. Goldberg, & F. Lobo (1999). A survey of optimization by building and using probabilistic models. Technical Report IlliGAL-99018, Illinois Genetic Algorithms Laboratory.

Prusinkiewicz, P. & J. Hanan (1989). *Lindenmayer Systems, Fractals, and Plants*, Volume 79 of *Lecture Notes in Biomathematics*. Springer, New York.

Prusinkiewicz, P. & A. Lindenmayer (1990). *The Algorithmic Beauty of Plants*. Springer, New York.

Rahman, A. & M. Fairhurst (1999). Serial combination of multiple experts: A unified evaluation. *Pattern Analysis & Applications* **2**, 292–311.

Rechenberg, I. (1994). *Evolutionsstrategie '94*. Stuttgart: Friedrich Frommann Holzboog Verlag.

Reidys, C. M. & P. F. Stadler (2002). Combinatorial landscapes. *SIAM Review* **44**, 3–54.

Rice, S. H. (1998). The evolution of canalization and the breaking of von Bear's laws: Modeling the evolution of development with epistatis. *Evolution* **52**, 647–656.

Rice, S. H. (2000). The evolution of developmental interactions: Epistasis, canalization, and integration. In J. Wolf, E. B. III, & M. Wade (Eds.), *Epistasis and the evolutionary process*, pp. 82–98. Oxford University Press, New York.

Riedl, R. (1977). A systems-analytical approach to macro-evolutionary phenomena. *Quarterly Review of Biology* **52**, 351–370.

Rissanen, J. (1978). Modelling by shortest data description. *Automatica* **14**, 465–471.

Ronco, E., H. Gollee, & P. Gawthrop (1997). Modular neural networks and self-decomposition. Technical Report CSC-96012, Center for System and Control, University of Glasgow.

Schuhmacher, C., M. Vose, & L. Whitley (2001). The no free lunch and description length. In L. Spector, E. Goodman, A. Wu, W. Langdon, H.-M. Voigt, M. Gen, S. Sen, M. Dorigo, S. Pezeshk, M. Garzon, & E. Burke (Eds.), *Genetic and Evolutionary Computation Conf. (GECCO 2001)*, pp. 565–570. Morgan Kaufmann.

Schuster, P. (1996). Landscapes and molecular evolution. *Physica D* **107**, 331–363.

Schuster, P. & K. Sigmund (1982). Vom Makromolekül zur primitiven Zelle – Das Prinzip der frühen Evolution. In W. Hoppe, W. Lohman, H. Markl, & H. Ziegler (Eds.), *Biophysik*, pp. 907–947. Springer, Berlin.

Schuurmans, D. (1997). A new metric-based approach to model selection. In *Proc. of Fourteenth National Conf. on Artificial Intelligence (AAAI 1997)*, pp. 552–558.

Schwefel, H.-P. (1995). *Evolution and Optimum Seeking*. John Wiley & Sons, New York.

Sendhoff, B. & M. Kreutz (1998). Evolutionary optimization of the structure of neural networks using recursive mapping as encoding. In *Artificial Neural Nets and Genetic Algorithms – Proc. of the 1997 Intern. Conference*, pp. 370–374.

Sendhoff, B., M. Kreutz, & W. von Seelen (1997). A condition for the genotype-phenotype mapping: Causality. In T. Bäck (Ed.), *Proc. of Seventh Int. Conf. on Genetic Algorithms (ICGA 1997)*. Morgan Kaufmann, San Francisco.

Shapiro, J. (2003). The sensitivity of PBIL to the learning rate, and how detailed balance can remove it. In C. Cotta, K. De Jong, R. Poli, & J. Rowe (Eds.), *Foundations of Genetic Algorithms 7 (FOGA VII)*. Morgan Kaufmann. In press.

Smith, J. & T. Fogarty (1997). Operator and parameter adaption in genetic algorithms. *Soft Computing* **1**, 81–87.

Stadler, B. M., P. F. Stadler, G. P. Wagner, & W. Fontana (2001). The topology of the possible: Formal spaces underlying patterns of evolutionary change. *Journal of Theoretical Biology* **213**, 241–274.

Stephens, C. & J. M. Vargas (2000). Effective fitness as an alternative paradigm for evolutionary computation I: General formalism. *Genetic Programming and Evolvable Machines* **1**, 363–378.

Stephens, C. & H. Waelbroeck (1999). Codon bias and mutability in HIV sequences. *Molecular Evolution* **48**, 390–397.

Sutton, R. & A. Barto (1998). *Reinforcement Learning*. MIT Press, Cambridge.

Toussaint, M. (2001). Self-adaptive exploration in evolutionary search. Technical Report IRINI-2001-05, Institut für Neuroinformatik, Ruhr-Universität Bochum, Germany.

Toussaint, M. (2002a). A neural model for multi-expert architectures. In *Proc. of Int. Joint Conf. on Neural Networks (IJCNN 2002)*, pp. 2755–2760.

Toussaint, M. (2002b). On model selection and the disability of neural networks to decompose tasks. In *Proc. of Int. Joint Conf. on Neural Networks (IJCNN 2002)*, pp. 245–250.

Toussaint, M. (2003a). Demonstrating the evolution of complex genetic representations: An evolution of artificial plants. In *2003 Genetic and Evolutionary Computation Conference (GECCO 2003)*, pp. 86–97.

Toussaint, M. (2003b). On the evolution of phenotypic exploration distributions. In C. Cotta, K. De Jong, R. Poli, & J. Rowe (Eds.), *Foundations of Genetic Algorithms 7 (FOGA VII)*, pp. 169–182. Morgan Kaufmann.

Toussaint, M. (2003c). The structure of evolutionary exploration: On crossover, buildings blocks, and Estimation-Of-Distribution algorithms. In *2003 Genetic and Evolutionary Computation Conference (GECCO 2003)*, pp. 1444–1456.

Toussaint, M. & C. Igel (2002). Neutrality: A necessity for self-adaptation. In *Proc. of IEEE Congress on Evolutionary Computation (CEC 2002)*, pp. 1354–1359.

Vapnik, V. N. (1995). *The nature of statistical learning theory*. Springer, New York.

Vose, M. D. (1999). *The Simple Genetic Algorithm*. MIT Press, Cambridge.

Wagner, A. (1996). Does evolutionary plasticity evolve? *Evolution* **50**, 1008–1023.

Wagner, G. P. & L. Altenberg (1996). Complex adaptations and the evolution of evolvability. *Evolution* **50**, 967–976.

Wagner, G. P., G. Booth, & H. Bagheri-Chaichian (1997). A population genetic theory of canalization. *Evolution* **51**, 329–347.

Wagner, G. P., M. D. Laubichler, & H. Bagheri-Chaichian (1998). Genetic measurement theory of epistatic effects. *Genetica* **102/103**, 569–580.

Wagner, G. P. & J. Mezey (2000). Modeling the evolution of genetic architecture: a continuum of alleles model with pairwise A×A epistatis. *Journal of Theoretical Biology* **203**, 163–175.

Watson, R. & J. Pollack (2002). A computational model of symbiotic composition in evolutionary transitions. *Biosystems, Special Issue on Evolvability* **69**, 187–209.

Wolpert, D. H. & W. G. Macready (1995). No free lunch theorems for search. Technical Report SFI-TR-05-010, Santa Fe Institute.

Wolpert, D. H. & W. G. Macready (1997). No free lunch theorems for optimization. *IEEE Transactions on Evolutionary Computation* **1**(1), 67–82.

Yao, X. (1999). Evolving artificial neural networks. *Proc. of IEEE* **87**, 1423–1447.

Zhigljavsky, A. (1991). *Theory of global random search*. Kluwer Academic Publishers.

Lebenslauf

Persöhnliche Daten

Familienname	Toussaint
Vornamen	Marc Alexander
Geboren am/in	12.02.1974 in Wertheim am Main

Bildungsweg

1980 – 1984	Grundschule Wertheim-Dertingen
1984 – 1991	Dietrich-Bonhoeffer-Gymnasium Wertheim
1991 – 1992	Pearland High School in Houston-Pearland, Texas
1992 – 1994	Dietrich-Bonhoeffer-Gymnasium in Wertheim
Juni 17, 1994	Allgemeine Hochschulreife
1994 – 1996	Grundstudium der Mathematik an der Universität zu Köln
Okt. 5, 1996	Vordiplom in Mathematik
1994 – 1999	Studium der Physik an der Universität zu Köln
Sept. 10, 1999	Diplom in Physik
seit Apr. 2000	Dotorand im Studienfach Physik an der Ruhr-Universität Bochum

Anstellungen

Mrz. 1998 – 2000	Studentische bzw. wissenschaftliche Hilfskraft am Institut für Theoretische Physik der Universität zu Köln
seit Apr. 2000	Wissenschaftlicher Mitarbeiter am Institut für Neuroinformatik der Ruhr-Universität Bochum